D0175212

Thirst for Power

Thirst for Power

Energy, Water, and Human Survival

Michael E. Webber

Yale
UNIVERSITY PRESS
New Haven and London

Published with assistance from the foundation established in memory of
Henry Weldon Barnes of the Class of 1882, Yale College.

Yale University Press books may be purchased in quantity for educational, business, or promotional use. For information, please e-mail sales.press@yale.edu (U.S. office) or sales@yaleup.co.uk (U.K. office).

Set in Bulmer type by Westchester Publishing Services, Danbury, Connecticut.
Printed in the United States of America.

ISBN 978-0-300-21246-4 (cloth: alk. paper)
Library of Congress Control Number: 2015950797
A catalogue record for this book is available from the British Library.

This paper meets the requirements of ANSI/NISO Z39.48-1992 (Permanence of Paper).

10 9 8 7 6 5 4 3 2 1

To Julia, Evelyn, David, Maverick, and future generations

Contents

Preface

This project has emerged over many years, with two key phases. The first was in 1990 during the spring semester of my freshman year in college, when I was studying the history of early civilization for one of my liberal arts classes. As an engineer just entering the world of liberal arts, I was a little out of place in the course's roundtable discussion format, but I found the topic fascinating and eagerly engaged in the debates. Partway through that year, in one of our classroom discussions, I concluded and stated that the most critical ingredients of modern civilization are water and energy. This idea stood in contrast with the prevailing hierarchy of food, water, shelter, and air, which had been posed as humanity's most important biological and physiological needs by psychologist Abraham Maslow in his famous 1943 paper "A Theory of Human Motivation."

Although Maslow's ranking seems true at the individual level, I thought that, at a collective level of civilization, the hierarchy might look very different: we need energy and water to make our food (water to grow the crops, energy to make the fertilizer and reap the crops), to make our shelters (water to grow the wood, energy to cut the wood down), and to achieve every other hallmark of highly modernized existence.

Years later, in 2005, while I was working at the RAND Corporation (a think tank) on some analytical projects related to energy, manufacturing, innovation, and security, my wife's aunt, Debbie Cook, who was serving on the city council of Huntington Beach, California (on her way to eventually serving as mayor), made a passing remark to me about how much water was used by the main power plant in her city. That remark brought my freshman realization rushing forward. This time, however, I was in a position to do something about it. I started drafting notes that turned into research projects, scientific articles, college lectures, speeches, book chapters, and op-ed pieces, ultimately creating a large body of work on whose lessons and stories this book is built.

In parallel and unbeknownst to me, Nobel laureate Rick Smalley of Rice University had created a list of "Top Ten Problems of Humanity for the Next 50 Years," which I discovered from Turk Pipkin's documentary *The Nobelity Project*. Smalley's list, organized in descending order of importance, was as follows: energy, water, food, environment, poverty, terrorism and war, disease, education, democracy, and population. That the top of my list coincided with that of a Nobel laureate was satisfying vindication and inspiration to pursue extensive research into the interrelationships between energy and water. I initiated the work while at the RAND Corporation, and expanded it significantly after joining the faculty of the University of Texas at Austin.

The order of importance in Smalley's list was not accidental. The challenges were stacked in sequence based on their overall importance to society. Notably, energy and water are at the top of the list, ahead of food and shelter. From Smalley's perspective, it is the fact that we can use energy and water to solve subsequent problems in a cascading fashion that puts them at the top. Energy is at the top of the list as the great enabler for all that follows. Developing abundant sources of clean, reliable, affordable energy leads to an abundance of clean water. Having an abundance of clean water enables food production and protects the environment, and so on down the list.

But there is more to the equation than Smalley's list demonstrates. It is not just the good news story that solving one challenge enables a solution for the other, but there is also the corollary that constraints in one can become constraints in the other. In fact, because of this interdependence, society is vulnerable to cascading failures in our infrastructure. Water shortages cause power outages, creating widespread disaster. Why did we design our society to be so vulnerable to these interconnections? How do we solve the problem, and what will our future look like when we do?

ONE

Healthy, Wealthy, and Free

WHEN I TALK ABOUT GEORGIA in the same sentence with an undemocratic land grab from a neighboring territory, people often assume I am referring to the country formerly part of the Soviet Union. But I'm not—water shortages can turn allies, even neighboring American states, into competitors. In 2008, the state of Georgia in the United States wanted land because it needed water: a year earlier various rivers dropped so low that the drought-stricken state was within a few weeks of shutting down its own nuclear plants. Water conditions had become so dire that the Georgia state legislature considered a resolution to move the state's upper border a mile farther north, cutting across the Nickajack Reservoir to annex freshwater resources in Tennessee. The justification for the boundary adjustment was an allegedly faulty border survey from 1818. That is, Georgia was trying to use a two-hundred-year-old map to execute a land grab to capture some water that otherwise was under the control of Tennessee. One Tennessee state senator, recognizing the importance of American football in the southeastern United States, jokingly suggested that the dispute should be settled with a football game between rival schools in the two states. That attempt failed, but water remains contentious for those states and their neighbors. Since then Georgia, Alabama, and Florida have continued to battle, with multiple lawsuits and allegations. Drought is only one cause. A rapidly growing population, especially in Atlanta, as well as overdevelopment and a notorious lack of water planning, is running the region's rivers dry. Production of thirsty energy sources just exacerbates the situation.[1]

But this isn't a crisis striking only close to home; it's global. In July 2012, the electric grid in India failed, causing the largest blackout in history.[2] It affected more than 620 million people, 9 percent of the world's population. Although there were many reasons for the power outage, it was a lack of water that triggered the Indian collapse. A major drought in India that summer increased the demand for electricity at the same time that it reduced electricity production. Because of the drought, farmers increased their irrigation of crops using electric pumps. Those pumps, working furiously under the hot sun, increased the demand for power, straining the grid. At the same time, low water levels meant that power generation at hydroelectric dams was lower than normal, making it doubly hard for the power sector to meet summer demand. Even worse, floods earlier in the year had caused dams to silt up with runoff from farms, reducing their available capacity even before the low water levels made things worse. A double water whammy of flood then drought hobbled the hydroelectric dams. The result? A population larger than all of Europe and twice as large as the United States was plunged into darkness, with railways and other critical services brought to a sudden halt.

Energy and water are the world's two most critical resources, but what many don't understand is even more significant: the two are intricately interconnected. They are like a hall of mirrors going on endlessly: energy needs water which needs energy which needs water . . . And strains in one can be crippling for the other with catastrophic consequences. In many ways, strains in the nexus of energy and water are our generation's Cold War—a global crisis spanning decades that we need to solve or the future of humanity will look very different.

Virtually everywhere we look, signs of the extent of this crisis abound. The same summer of the blackout in India, a massive heat wave and record-setting drought swept across most of the United States, putting power plants in peril because cooling water was scarce or too hot to be effective. A few years earlier, the state of Florida made an unusual announcement: it would sue the U.S. Army Corps of Engineers over the Corps' plan to reduce water flow from reservoirs in Georgia into the Apalachicola River, which runs through Florida from the Georgia-Alabama border.[3] Environmentalists were concerned that the restricted flow would threaten certain endangered species. Regulators in Alabama also objected, worried about another species: nuclear power plants, which use enormous quantities of water, usually drawn from rivers and lakes, for cooling. The reduced flow raised the specter that the Farley Nuclear Plant near Dothan, Alabama, would need to shut down.

In California, two of its fabled nuclear reactors are sited on the coast between Los Angeles and San Diego. As one drives along I-5 between the cities, these nuclear reactors look surprisingly like two gigantic concrete breasts rising above the coast, just beside the highway and silhouetted by the ocean. Despite their evocative shape and their appealing, innocent-sounding name—SONGS, which stands for San Onofre Nuclear Generating Station—SoCal Edison announced that these plants would be shut down, partly because of water. Ongoing concerns that the nuclear power plants put ocean life at risk from entrainment of cooling water and the potential for radiation leaks that could contaminate the water along with other technical challenges became too burdensome to manage. It was easier to shut down the plant than to solve the radiation leaks and water problems to the satisfaction of the public.

Outside Las Vegas, Lake Mead, fed by the Colorado River, is now routinely one hundred feet lower than historic levels. As it stands, Las Vegas draws its drinking water from two straws that tap into Lake Mead. If the lake drops another fifty feet, the city's water supply will be threatened and the huge hydroelectric turbines inside Hoover Dam on the lake would provide little or no power, potentially putting the booming desert metropolis in the dark while leaving its occupants thirsty. Las Vegas' solution is to spend nearly $1 billion on a third straw that goes deeper, coming from underneath and up into the lake from the bottom. But even that drastic solution might not work. Scientists at the Scripps Institution of Oceanography in La Jolla, California, have declared that Lake Mead could become dry by 2021 if the climate changes as expected and water users who depend on the Colorado River do not curtail their withdrawals.[4] We can argue about whether the glass if half-full or half-empty, but an engineer will point out that the length of your straw does not matter if the glass is dry.

Communities in water-strained Texas and New Mexico, wary of the water risks posed by hydraulic fracturing for oil and gas production from shale formations, have imposed prohibitions or constraints on what water (if any) could be used for fuel extraction, even though the amount of water required by shale production is small compared with agriculture. Activists opposed to oil and gas production use concerns about possible water pollution as the rallying cry to halt increases in drilling.

Energy can be a limiting factor for water, too. San Diego, which needs more drinking water because of its growth and the pervasive California droughts, has sought to build a desalination plant on the coast. But local activists have fought the facility—it would consume so much energy, and the power supply is thin. For the same reason, the mayor of London denied a proposed desalination plant in 2005, only to have his successor later rescind that denial.[5] Politicians in Uruguay must choose whether they want the water in their reservoirs to be used for drinking or for electricity. Saudi Arabian leaders wrestle with whether to export the country's oil and gas to earn hard currency or to burn more of those resources at home to produce what it does not have: freshwater for its people and its cities.

When Hurricane Katrina struck New Orleans in August 2005, the destruction was awe-inspiring. Amazingly, the city was peaceful after the hurricane passed . . . at least initially. But the widespread power outage meant that the water system—including pumps to keep low-lying areas dry and the treatment facilities that clean the water to potable standards—quit working. It was only when people realized that their drinking water was contaminated and that their neighborhoods were not going to be pumped dry anytime soon that chaos broke out. An energy shortage sparked a water shortage, destabilizing society.

We cannot build more power plants with the same old design or extract more oil and gas with outdated techniques without realizing that they impinge on our freshwater supplies. And we cannot build more water delivery and cleaning facilities the same way we always have without driving up energy demand. If we continue in the direction we're going, widespread vulnerabilities in our interconnected water and energy systems will worsen with population growth, economic growth, and climate change, all of which exacerbate the strain. Sadly, we often compound the problems with self-inflicted policy decisions that push us toward more energy-intensive water and more water-intensive energy.

Despite the importance of each and the close relationship between energy and water, the funding, policymaking, and oversight of these resources are typically performed by different people in separate agencies. Energy planners often assume they will have the water they need and water planners assume they will have the energy they need. If one of these assumptions fails, the consequences can be dramatic, as the blackouts in India demonstrated.

But we can stop this downward spiral if we seize all the potential advantages to the nexus of energy and water. With abundant, clean, reliable, affordable energy we can solve our water problems by desalting oceans, digging deeper wells, and moving water thousands of miles uphill to thirsty people and crops. And with abundant, clean, reliable, affordable water, we can solve our energy problems by building hydroelectric systems to generate all the electricity we could ever desire, and we could grow our way out of our oil problem by irrigating biofuels. But because we don't have infinite resources, we're dealing with a world of constraints instead. While the challenges are difficult and the potential risks are great, with new thinking and some clever innovation we can manage this problem for a better future.

Solving the dilemma requires new policies that integrate energy and water solutions and innovative technologies that help to boost one resource without draining the other. Thankfully, we have technical solutions that make sense. There are water-lean energy options and energy-lean water options that we could implement. But we do not typically select them because the world's politicians and decision-makers have not fully grasped the interrelatedness of these resources, and policymakers, as well as engineers, are isolated and confined to either water or energy systems.

There's potential for more good news: because we use so much energy for water and so much water for energy, we have the opportunity for cross-cutting conservation. By saving water, we can save energy, and by saving energy, we can save water. Most of us don't realize that we use more water for our light switches and electrical outlets than our faucets and showerheads because the water is used to cool power plants far away, out of sight and out of mind. It also doesn't occur to us that as a nation we use more energy to heat, treat, and pump our water than we use for lighting. Counterintuitive to what we'd expect, turning off the lights and appliances saves vast amounts of water, and turning off the water saves vast amounts of energy.

In the end, the most important innovation we need is a new way of thinking about energy and water so that we make better decisions about these precious resources: holistic thinking that recognizes these resources as interconnected, and a systems-level approach that acknowledges how one change in one state to a water system could impact an energy system five states away. Most important, we need long-range thinking because our energy and water decisions last decades to centuries, so it's imperative that we get them right. This book will show us how we can change our thinking about energy and water to be more integrated, with the goal of long-term sustainability, and that, once we adopt this new mindset, many solutions open up that will enable us to manage the water-energy nexus holistically and set us up for a better future.

Water, energy, and civilization go hand in hand. The various multicentury Chinese empires survived as long as they did in part by controlling floods in the Yellow River.[6] This political and imperial power is captured in the word *zhi*, which has simultaneous meanings "to rule" and "to regulate water." In fact, an article by the *Economist* in 2009 noted that "the Chinese word for politics (zhengzhi) includes a character that looks like three drops of water next to a platform or dyke. Politics and water control, the Chinese character implies, are intimately linked." Indeed, water and politics go hand in hand for many societies and cultures, not only the Chinese. In the social sciences, there's a hydraulic theory of civilization in which water is the unifying context and justification for many large-scale civilizations, and we can see it playing out in a variety of contexts throughout history. One interpretation of this idea is that the justification for forming large cities in the first place is to manage water, and that large water projects enabled the rise of megacities; cities and water projects go together.

The Romans certainly understood the connections between water and power: they built a vast network of aqueducts throughout their empire, many of which are still standing. The aqueduct of Segovia in Spain was operational after nearly two thousand years up until the twentieth century. And, the magnificent Pont du Gard in southern France proudly stands as a testament to humanity's investment in its water infrastructure.

According to Trevor Hodge, whose book on Roman aqueducts is frequently cited, "The aqueducts went wherever Rome went, an outwards symbol of all that Rome stood for and all that Rome had to offer."[7] In other words, the Romans would build roads, bridges, and water systems as a way of Romanizing new territory. Ultimately, the Roman empire's vast water infrastructure came to be considered one of the ancient world's greatest achievements. And, similar to the Chinese empires, it helped them keep a hold on their power, much like modern politicians of today who erect dams as monuments to themselves.

Other ancient civilizations collaborated to build massive waterworks, like the Khmer empire's vast water network that reached its apex in the thirteenth century. The most famous feature of this water system is the temple known as Angkor Wat.[8] That complex included several water retention ponds, storage systems, distribution channels, and a water temple, which is the main religious shrine that most people associate with the site. And those water temples are not just something that far-flung ancient civilizations built in the middle of jungles: the San Francisco Bay Area also has a modern-day water temple that was erected in 1934 to celebrate the arrival of piped water from the Hetch Hetchy reservoir over 160 miles away in the Sierra Nevada mountains.[9]

Just as water and civilization go hand in hand, so too do sustained water scarcity and societal downfall. We've seen this play out in the surprising number of civilizations that have collapsed over history, often with ecological strain as precursors—in particular sustained drought and ensuing stress on water systems.[10] While Angkor Wat and the surrounding water system were strategic assets that helped consolidate power, when the water system failed, Angkor's power went with it. Other research has revealed that three of the five multicentury Chinese dynasties—the Tang (618–907), Yuan (1271–1368), and Ming (1368–1644)—collapsed at times coincident with multidecade drought.[11] That research examined a stalagmite in the Wanxiang Cave, China, at the northern fringes of the monsoons and therefore a useful indicator of strong or weak monsoon activity. The stalagmite that provided the key historical timestamp formed slowly over 1,810 years starting in the year 190 CE, creating a finely resolved year-over-year mineral record that reveals how wet or dry that particular year was. Lining up those mineral records with Chinese written records, which go back thousands of years, showed a remarkable sequence of eras with major societal expansions and collapses or civil unrest. During wet years, when there was enough water for rice cultivation, populations expanded. For example, the Northern Song dynasty (960–1127) doubled in population, going from about 50 million people to more than 100 million in the span of 150 years. During extended periods of weak monsoon activity, dynasties struggled and fell. The coincidence in timing is striking.

Research in caves in Israel produced a similar conclusion: the Roman empire, whose vast water system with sprawling aqueducts was one of its greatest achievements, also collapsed at a time coincident with sustained drought.[12] Although the invading hordes of tribes is one of the conventional explanations for the empire's demise, I cannot help but wonder if water strain at the edges of the Roman frontier might have driven the tribes toward invasion.

Other water-related societal failures include the Mayan collapse around 900 CE, which is one of the more spectacular collapses in recorded history. Whereas debate remains about the precise timeline and causes of the Mayan collapse, researchers have suggested that climate change and extended drought along with failure of Mayan water systems triggered a contraction of their population by 50 percent or more over the span of decades.[13] The climate record from a cave in Belize indicates that there was a rise in Mayan warfare when drought was under way, suggesting that water wars were a contributing factor. Interestingly, the Mayan downfall happened at roughly the same time as the Tang dynasty was ending in 907 CE. Brian Fagan's work on the rise and fall of civilizations ties these two events together with global climate change.[14] Later on, drought hit the Khmer empire and the Anasazi Indians of the American desert Southwest at around the time those empires collapsed. But it was not bad news everywhere. Europe had the problem that its water abundance caused crops to rot and mildew. Lower rainfall improved agricultural yields and multigenerational political stability in Europe. That means while civilizations elsewhere were imploding, Europe was thriving, building projects over multiple generations such as massive cathedrals that still stand today.

In addition to the dramatic impacts of water availability and scarcity on civilization, climate changes affect the global water cycle, creating a cascade effect in which climate change ultimately means civilization change. Although the concerns over climate change are multifaceted—its worst effects include declining crop yields, increasing ocean acidification, and decreasing comfort in many parts of the world—the changes in water are perhaps the most critical as those are what provide the largest potential impact to society. In an unfortunate feedback loop, global climate change impacts the hydrologic cycle by increasing the frequency and intensity of droughts and floods. Those events drive increases to our energy consumption as we use energy to mitigate problems with the water system. Consuming more energy emits more carbon, which drives the whole process faster in a pattern that works against us.

Other examples abound. The regions in the Middle East and North Africa torn by civil unrest today also seem to be the ones where there is a battle over access to energy resources or where drought has strained food supplies.[15] Drought in Syria bankrupted farmers, who fled to the cities to find work. Large urban populations of unemployed farmers, along with poor governance, fanned the flames of civil unrest, ultimately triggering a refugee crisis.

Because modern water systems depend so heavily on energy, modern society depends on energy and water. Unfortunately, an energy shortage can become a water shortage, critically destabilizing society. It wouldn't be an exaggeration to say that the fate of civilization depends on the comingling of our energy and water systems, and how we solve the crises these systems currently face.

The quantity of water is not the only factor that matters to the fate and relative prosperity of a society: the water must also be clean. Despite the importance of clean water and sanitation, the close connections between public health and water supplies were not revealed scientifically until the mid-1800s. The first scientific identification that cholera is spread by water sources contaminated with human waste was made in 1849 by Dr. John Snow in London.[16] As Bill Bryson explores in *At Home: A Short History of Private Life,* Snow's findings were produced between two famous cholera outbreaks that struck London in 1848 and 1854. After the 1848 outbreak, Snow was able to determine that public wells that drew water from the heavily contaminated Thames (at the time, untreated sewage was emptied into the river) were the source of the problem.

The sudden spike in mortality exceeded that of London's famous plague episodes, producing such spectacular consequences as five hundred people dying in a small neighborhood over the span of just ten days. Unfortunately, Dr. Snow's findings that human waste was contaminating water and killing people were rejected by Parliament because they did not fit prevailing ideologies and because the actions that would be required to fix the problem were deemed too expensive. Similar rejection is offered for today's climate scientists, who tell us our waste is killing us, though in a much slower and less direct pathway, and that fixing the problem will require significant investments in new infrastructure. Snow was later vindicated as a hero, and perhaps the same fate awaits our present-day scientists who warn us of climate risks and offer solutions.

Bryson reveals another telling anecdote about an episode known as "The Great Stink" in London in 1858. Although untreated sewage had been dumped into the Thames for decades, the currents had done a great service to the metropolis by washing the waste away to sea. However, the combination of heat wave and drought that summer meant there was less water to dispose of the wastes. That stagnant water and stagnant air produced a remarkably noxious smell, triggering a temporary suspension of

Parliament, whose building sits within smelling distance of the Thames. While the event was unfortunate and unpleasant in many ways, it led directly to the creation of an ambitious public works project to insert twelve hundred miles of sewers into a crowded city of 3 million people. Doing so not only addressed the problem of waste disposal, but it also created the lovely river embankments that still stand today as a key piece of London's urban landscape along which many people stroll.

In the mid-1800s, Londoners could solve their water problems by simply flushing waste away farther along the Thames. But, today, with a higher global population and density, there is no "away." It's impossible to rely on dilution as the solution. Instead, industrialized societies invest energy: energy for water treatment and energy for wastewater treatment. And, like the London experience, where the sewers became walkways, if we do it the right way, we can solve our water problems while simultaneously building structures we can use for other purposes. Investing energy to clean our water is one of the great public policy achievements for modern civilization in the past 150 years.[17] Energy also lets us heat our water, which is critical for sterilizing medical equipment, washing our hands, ridding our society of many disease-carrying pests (many pesticides are made from petrochemicals), and cleaning scrapes and wounds.

Although the scientific and democratic advances since the industrial revolution have been significant, the largest public health problem globally remains the more than 1.1 billion people without access to clean water sources for drinking, cooking, and washing. That number is expected to grow to 1.8 billion people by 2025. In China alone, 100 million people lack improved water sources, and 2.6 billion globally remain vulnerable to waterborne diseases because they lack access to sanitation (which is a polite word for wastewater treatment).[18] Nearly 4.8 billion people, or 80 percent of the world's population in 2000, reside in areas with significant water security or biodiversity threats. Improving water quality is a significant way to improve public health worldwide.

Delivering universal access to clean water to improve public health will require a lot of energy for treatment and transport to where it is needed. And if we're not careful, just throwing more energy at the problem might make other problems worse. Even though the energy can be used to clean up the water and improve public health, if we use energy that pollutes, then it might undo those benefits by depositing pollutants back into the water systems. Air laced with pollutants from smokestacks and tailpipes can cause premature mortality or weaken productivity because of sick days that keep employees from work, either because they do not feel well, or because they have to stay home to tend to a sick child. Emissions from our energy consumption accumulate in the air. With the right meteorological conditions, those emissions are converted into dangerous chemicals, such as ozone, or fine particles that settle deep into our lungs and cross into our bloodstreams. If the concentrations of these substances are too high, then the air actually becomes dangerous to breathe. "Ozone Action Days" are declared on the days when ozone levels are high enough to cause damage, and outdoor physical activity is discouraged.

Poor air quality is causing an asthma epidemic in the United States, and it has been declared one of the most important public health problems worldwide, afflicting millions of people.[19] Asthma attacks alone cause $20 billion in health expenses annually in the United States, and the epidemic is growing in its severity. Far more deadly than auto accidents, air pollution is blamed for 7 million premature deaths each year worldwide, more than 4 million of which are from household air pollution due to activities such as burning fuels like wood or cow dung in primitive cookstoves that produce a lot of smoke and soot. Air quality problems cause $150 billion in economic losses and 150,000 to 200,000 deaths in the United States alone. And that's despite the relative cleanliness of U.S. air compared with other energy-intensive countries.

Coal is one of the main culprits for bad air quality. In China, where coal is used for generating two-thirds of the country's electrical power, often in dirty, unscrubbed power plants, the air pollution is breathtaking—literally. The numbers are staggering: at least 1 million people die annually in China from air pollution.[20] In the United States, where we use less coal than China, and where the coal itself is cleaner (for the most part, U.S. power plants use low-sulfur coal, which has fewer pollutants), and where it is burned more cleanly (U.S. power plants are more likely to have scrubbers than Chinese power plants), coal combustion still causes expensive damage to our public health and ecosystems.

Looking at coal through the lens of the energy-water nexus gets tricky. The same coal that dirties our air and water also makes electricity that we use to clean our water. And an electric cookstove powered by electricity from a coal-fired power plant far away is cleaner than burning cow dung in our homes. How are we going to balance this tension in the future so that we continue to have the clean water and stoves, but without all the pollution?

Ashlynn Stillwell, a professor at the University of Illinois at Urbana-Champaign, pointed out to me that the unit of currency in Botswana is the "pula," which literally translates as "rain." In the movie *Rango,* the municipal bank in the middle of a desert community stores water in the vault, not gold. Terms like petrodollars and black gold suggest that energy and money are also synonymous in some contexts.

As the building blocks of industrial processes and agricultural production, energy and water both foster wealth creation and prosperity. And the consumption of energy and water also increases with wealth. Rich people consume a lot of energy for their homes, cars, and food, and use a lot of water for their lawns. As with many correlations, it is not clear if affluence causes energy and water consumption, or if consumption causes affluence, but the relationship is salient.

Whereas the United States is responsible for approximately 20 percent of global energy consumption, it is also responsible for about 20 percent of global economic activity. In fact, it turns out that energy consumption and economic activity have a roughly linear relationship. That is, countries that have higher per capita energy consumption tend to have higher gross domestic product. Per capita energy consumption is dependent on the wealth, lifestyle, culture, prevailing climate, and affordable access to energy for each country. It ranges from more than 800 million British thermal units (Btu) of energy per person per year for a small, rich, resource-abundant country such as Qatar to less than 10 million Btu per person per year for residents of Eritrea, a poor country with limited access to resources.[21]

This type of consumption has a positive feedback loop: countries with a lot of energy resources can become rich either by extracting those resources to sell to others or by harnessing them directly for their own economic activity such as manufacturing or agriculture. Whether those resources enrich the entire population or a small subset depends on factors such as effective governance structures and market systems. Then, as people become wealthier, they tend to consume more energy for electricity, meat, and transportation, all of which are typically preferred by the affluent. The water and energy used to make those products are called embedded energy and embedded water. Some call it the energy footprint, water footprint, or virtual water. I consume more water embedded in the grains I eat because of their irrigation than from the faucets at my house.

For a while, this pattern is a virtuous cycle, but at some point the cycle reverses itself because of the negative health effects of consuming dirty energy sources: energy consumption begets wealth, which begets more energy consumption, which begets pollution, which reduces wealth. We need to clean up our energy systems so that our consumption makes us healthy and wealthy instead of sick and poor.

As energy consumption varies by national income, so does water use. Affluent people eat more meat, which is very water intensive, because of all the water used to grow the feed for the livestock. The rich also use more electricity for appliances and tools. Electricity is also very water intensive, since power plants use a lot of water for cooling.

Not only do poor countries withdraw less water per person, but they also withdraw much less for industrial uses. Poor countries use a much higher fraction of their water for agriculture because they are closer to subsistence farmers, using most of their water just to feed themselves. By contrast, for the industrialized nations, where food needs have mostly been met, water can be used for other purposes. In the United States, only 39 percent of water withdrawals are for agriculture, compared with 70 percent globally.[22]

Mary Clayton was a physically fit and active twenty-two-year-old student at the University of Texas when she spent a month during the summer of

2011 doing project work in Ghana to help build a water system for a school. In Ghana, as in many parts of the developing world, it is the responsibility of women to fetch the water. Clayton came back describing the enormous burden of carrying water and what a challenge it was for her. She also shared anecdotes of girls—including girls younger than school age—who were stronger and able to carry more water a greater distance than she was. She fondly remembers the young girls laughing while trying to help her lift the bucket to her head that they so easily carried every day. If it is a burden for a fully grown, strong, healthy woman, imagine how many times girls in Ghana must have carried that burden to build the strength they needed.

Those girls would miss hours of school each day to get water from far away, carrying the heavy jugs of water balanced on their heads to cover a distance over a mile between the well and the school. But once they have the water, they are not done, as the water needs to be treated before it can be drunk. In remote villages where piped water systems and centralized water treatment with electrical pumps and other advanced techniques are not available, water is treated the old-fashioned way: it must be boiled. There's a similar story for getting fuels: women often have to collect fuel from remote areas, again vulnerable along the way.

When they return after having fetched water and fuels, they need to use the fuels to boil the water as a form of treatment. Those fuels—including crop residues, animal waste such as cow dung, wood, and untreated coal—are burned in inefficient and dirty stoves that are used for cooking and heating. Unfortunately, the outdated cookstoves perform badly, producing indoor air pollution that has been linked to the premature death of over 4 million people every year, more than half of them women and children.[23] In other words, antiquated energy and water systems put women at risk when they are collecting the water and the fuels, and when they are using the fuels to treat the water. These old energy and water systems literally deprive girls of their education and kill women by the millions.

Such archaic, labor-intensive approaches take a toll on prosperity and economic opportunity.[24] In many places, the world's poorest women are also traditionally responsible for planting and harvesting crops, milling grain, and fulfilling household chores; tedious responsibilities that leave little time for an education or employment outside of the home. Even if they could go to school, they might not have the lights they need to read books at night to study or, as the earlier anecdote noted, have to miss hours of school to fetch water. In turn, they have few or no options to work, earn an income, and gain independence, which perpetuates poverty. This fact is especially galling considering that more than 70 percent of the 1.5 billion people living on less than one dollar a day are women.[25]

Although the scenario I described sounds like something from the developing world somewhere far away, it is also part of the United States experience in the not-too-distant past. The University of Missouri in 1920 issued a poster as part of an information campaign—or propaganda, depending on whom you ask—to encourage farm owners to modernize the water systems of their homes and operations. It is aptly titled "The Farm Woman's Dream," and shows a woman—presumably poor and dressed in rough clothes or rags—carrying water without any gloves in the freezing cold along an icy path uphill from a hand-pumped well. The dreary image evokes a sense of hard labor associated with getting water into our homes. In the upper part of the image is the farm woman's purported dream: nicely dressed in short sleeves, comfortably indoors, opening spigots at a sink, with hot water coming out, its steam curling to the ceiling. The poster is targeted at women, not men, a telling example of who endured the greatest burden from not having access to modernized systems, and who would reap the greatest benefit from the modern improvements. The burdened, vulnerable woman in Africa today, struggling to fetch water, is not much different than the burdened, vulnerable woman in the rural United States less than a century ago. Access to water and energy turns the story around.

For that poor hill country woman and the vulnerable women in Africa today, their dream—the solution—is the same: with a modern water system (with pipes and pumps) and a modern energy system, including electricity to drive the pumps and fuels to heat the water, women can be liberated from tedious and dangerous chores. Novel electrical appliances such as the dishwasher and washing machine provided women in the United States even more freedom to pursue opportunities outside the home beyond menial labor.[26] Chores like cooking and cleaning were no longer

"The Farm Woman's Dream": A U.S. government poster from the first part of the twentieth century depicts piped heated water indoors as the dream of every farm woman. [University of Missouri, College of Agriculture, Special Collections, National Agricultural Library, 1920]

as time consuming and complicated, and a new generation suddenly had the chance to pursue higher education and their own careers. In turn, American culture began to shift to accommodate working women.

Similar changes will take place in the developing world when we work harder to promote greater access to electricity and the adoption of modern energy tools. Distributed energy and water systems that use advanced technologies like nanofilters and smart controls can help democratize resource access, improve health, and liberate a whole new generation of people. Improved living conditions from cleaner and more reliable energy and water services would open up the possibility for alleviating poverty among women while also giving them more choices for goals in life.

Typically, international policy discussions regarding energy and water specifically focus on economics, security, and the environment, leaving human rights and women out of the conversation entirely. But by working to improve global access to more efficient sources and making these sectors more resilient, we will achieve a healthier, cleaner, and more sustainable future for all of us. Doing so will empower women, which in turn empowers society.

TWO

Energy

We have no knowledge of what energy is. . . . However, there are formulas for calculating some numerical quantity. . . . It is an abstract thing in that it does not tell us the mechanism or the reasons for the various formulas.

—Richard Feynman, *The Feynman Lectures on Physics,* 1963

ENERGY IS THE FUNDAMENTAL driver of most of the processes we care about. It helps drive photosynthesis to make plants grow, it drives our cellular functions, and its modern forms are what make today's societies different than those of antiquity. Energy even has its own language, with nuanced differences in the meaning of different vocabulary terms that can be confusing.

Energy is the ability to do work, where work is defined as exerting force over a distance. Lifting a rock and pushing a wheelbarrow are examples of work. While energy can be evaluated, predicted, and controlled, it remains an abstract concept that is hard to define, touch, or describe. Despite the fact that energy surrounds us—and is embedded within us—in many ways, it is hard to say what energy is. Even the Nobel laureate physicist Richard Feynman, who worked deeply with quantum mechanics and electrodynamics, and who is credited with introducing core concepts of nanotechnology, pointed out that we don't really know what energy is.

Power is closely related to energy, but is different. While energy is a quantity, power is a rate. It is the rate at which energy is produced, moved, or consumed. That means power is a measure of energy per unit of time, or work per unit of time. Power measures how quickly something is being done or how quickly energy is being consumed.

As a scientist, engineer, and thermodynamicist, I find it is easier to describe the consequences and transformations of energy than it is to describe the fundamental nature of energy itself. And this ambiguity is just one of energy's many challenges. The definition I use in the classroom while teaching thermodynamics is that energy is something we use to predict and explain how things happen. I can teach you what energy does. I can explain how to harness it, convert it, manipulate it, and clean up after it. But we cannot touch it or see it. Yet the evidence of its existence is all around us, omnipresent—and maybe omniscient, too, if you include the information contained within energy—like a mysterious force.

We owe the modern definition of energy to the nineteenth-century physicist and inventor Sir Benjamin Thompson. He challenged established physical theory and spurred a revolution in thermodynamics, which is the underlying science of energy. Just the word "thermodynamics" itself is telling. *Thermo* means heat (like a thermos that keeps your soup warm, thermal underwear that keeps the heat close to your body, or a thermometer that measures the heat), which is a proxy for energy. And *dynamics* means change (or changes), which is the opposite of static or steady. So, the field of thermodynamics is the study of changes in heat or changes in energy. Thermodynamicists study those changes, to understand how energy works and how to harness it better.[1]

Energy's relevance is based on transformations from fuels such as coal, oil, wood, or gas to motion that allow us to perform useful functions such as mechanical work or moving a car or crushing rock. Or we can change the chemical energy in the fuels to heat for thermal activities such as warming and cooking. While providing useful services, these transformations, which are governed by the laws of thermodynamics, also cause pollution and waste.

As a professor who gives many lectures for the general public and teaches in many disciplines across campus for experience levels ranging from freshmen through Ph.D. students, I am often in position to give introductory lectures on energy to nonengineers. Most nonengineers groan when I warn them that I am about to give a lecture on thermodynamics. I usually reply that thermodynamics is like Shakespeare for engineers. And, since nonengineers make us engineers take classes on Shakespeare whether we want them or not, it seems only fair to return the favor.

For that matter, engineers often groan, too. Thermodynamics, which is the science at the heart of energy, is usually offered during the sophomore year of college and is considered one of the last weed-out classes for most major engineering programs at top-tier universities. The class is notoriously hard. Interestingly enough, despite the difficulty of the subject, it is not unusual for engineering students many years removed from the class to fondly recall thermodynamics as the first course when they actually felt like they understood how the world works.

The topic is important because the laws of thermodynamics provide the key principles by which energy is harnessed or wasted. That is, there are physical laws that describe changes in energy. The typical joke is that when hearing about the laws of thermodynamics for the first time, and hearing that the laws have some undesired outcomes for humanity, a savvy politician suggested that legislation could be passed to change the laws and a high-dollar lawyer started looking for loopholes in the law. Unfortunately (or fortunately?), such a legislative option is not available and no loopholes exist, as these physical laws are immutable, as dictated by nature.

Compared with the teachings of philosophy or the emergence of the world's major religions, the knowledge of these laws came about only relatively recently. The science of thermodynamics developed in the 1800s from the desire to improve the performance of early steam engines. These engines were increasingly valuable in Britain just after the Industrial Revolution as they provided mechanical power at factories and mills.

As if to illustrate that the nexus of energy and water has been important since the dawn of the Industrial Revolution, one of the most important early applications of steam engines was to pump water out of coal mines. As the coal near the surface was extracted first, miners eventually had to dig deeper to find productive coal seams. As mines went deeper into the ground, water would pool at the bottom of the pit and get in the way of the miners. Pumping the water out of the mines took tremendous effort, so steam engines were used to do the heavy lifting. With a seemingly circular loop of the energy-water-energy-water-energy nexus, energy in the form of coal was used to drive a steam engine (steam is a form of water) that was used to provide mechanical energy to pump water so more coal could be mined. Notably, some coal mines today still need to be dewatered.

Most of the early work inventing steam engines was accomplished with the slow march of progress from trial and error. Early tinkerers needed to understand the theory of the engines to make them work better, so they needed new science. The names of some of those early scientists—Thompson, Carnot, Watt, and Joule—are still recognizable today and some are immortalized as units of power and energy.

As prominent science historian Bruce Hunt noted, it is not clear who helped whom the most: while we often think that innovation follows a path from new scientific discovery in the lab leading to better engineering design in the field leading to a better product available for consumers, the history of thermodynamics might have gone the other way.[2] Rather than going from fundamental science to applied science to prototype to product, the direction was reversed. That is, the tinkerers inventing and improving engines out at the mines and in the factories might have done more to advance the science of thermodynamics than the thermodynamicists might have done to improve the engines.

There's the old adage among engineers that asks, "We know it works in theory, but does it work in practice?" For the early study of the science of energy, it might have gone the other way around: "We know it works in practice, but does it work in theory?" The engineers were able to figure out how to convert heat into motion before the scientific theory existed to explain what was going on. But today, we know that what's happening with energy is governed by three laws of thermodynamics.

1. The best you can do with energy systems is break even.
2. You are not going to break even.
3. Energy processes never stop.

Of these three laws, the first two are most important in the context of our energy future. The First Law of Thermodynamics is one of the fundamental governing laws of the physical universe, and includes three separate but related concepts:

1. Energy is conserved.
2. Energy has many forms.
3. Energy can be converted from one form to another.

The most important implication of the First Law of Thermodynamics is that energy is conserved. That is, we can never get more energy out of a system than what is put into it. If a system produces more energy than is put into it, then that system violates the First Law of Thermodynamics.

A quick search of the Internet will reveal all sorts of innovative designs for machines that their inventors claim will provide perpetual motion. That is, these machines, once started, never need fuel or additional inputs to propagate their motion. According to the First Law of Thermodynamics—and practical judgment, too—such machines are impossible. So save your money. That is, one cannot get energy for free—it has to be paid for in some way.

The First Law of Thermodynamics simply states that all energy is conserved. No matter what we do, in a closed system, energy will always be conserved; it is neither created nor destroyed. It can only be changed from one form to another or transferred from one body to another. The total amount of energy remains constant. Another way to consider this concept is that the best one can ever do with a closed system is break even: there will never be more energy than what it started with. Energy cannot be earned for free; it has to come from somewhere else.

While the First Law of Thermodynamics says that energy must be conserved, it also says that energy can exist in different forms. The different forms of energy are sometimes in raw form found in nature, such as the chemical energy stored in the bonds of molecules of fuels such as petroleum, coal, wood, or natural gas. Sometimes it exists in a more useful form such as directed radiant energy—think of the energy in a laser beam or spot lighting from incandescent bulbs—or mechanical energy of a moving object. The typical forms of energy include chemical (c), atomic (a), electrical (e), mechanical (m), radiant (r), and thermal (t).

While energy is conserved, and energy has many forms, what makes the First Law of Thermodynamics particularly useful is that we can convert energy from one form that is convenient for storage—for example, bound up as chemical energy in a cord of firewood—into another form that is useful for our comfort—for example, as thermal energy emanating from a fireplace that warms our house. In fact, the intentional and thoughtful transformation of energy from one form to another is what enables many of the great aspects of modern society: physical mobility, climate control, refrigeration, and so forth. Manipulating these forms of energy is one of the behaviors that distinguish humans from other species.[3] While many animals uses muscle power to break open shells or to crack nuts, no other species controls fire or manipulates chemical energy in such an intentional way.

While this ability to convert energy has no end of practical uses, it is still bound by the requirement that energy is conserved. In practical terms, whenever a fuel is transformed—for example by burning a cord of wood in a fireplace—the magnitude of energy contained in the heat, light, and waste that is produced by the burning will be the same as the original chemical energy contained within the unburned wood.

To illustrate the concept, consider the process of lighting an incandescent bulb using coal-fired electricity. While flipping a light switch seems so simple, that action triggers a sequence of four energy conversions through five different forms of energy. This process begins with chemical energy (c) in the form of coal, and ends with radiant energy (r) in the form of light from the bulb.

To begin, the chemical energy in coal (c) is converted to heat or thermal energy (t) when the coal is burned. A boiler uses the heat to convert liquid water into steam that drives a turbine, similar to the way steam shooting out of a tea kettle would spin a pinwheel if you were so inclined to put one in front of the spout, giving mechanical energy (m) from its rotation. The turbine spins electricity generators that rotate magnets giving electrical energy (e), which is then distributed to the home or building where the bulb will be used. The incandescent lightbulb converts the electricity to radiant energy (r) in the form of light. The overall process goes from chemical to radiant energy ($c \rightarrow r$) through a series of individual sequential conversions ($c \rightarrow t \rightarrow m \rightarrow e \rightarrow r$).

But something happens along the way: there are losses and inefficiencies and waste heat. What causes those losses? The Second Law of Thermodynamics: it prevents our processes from breaking even.

The Second Law of Thermodynamics dictates that entropy (or disorder) increases for a closed system. For example, imagine a dorm room on the first day of college. The room starts off neatly ordered. Flash forward a few weeks and the room is highly disordered, with books, clothes, and snacks scattered about. That familiar anecdote is an example of disorder increasing with time. Left to their own devices, dorm rooms and their inhabitants naturally go from clean to messy. Scientifically, this requirement describes the direction of different processes. Consequently, some people refer to entropy as the arrow of time.

For another example, imagine a beautiful maple tree in the autumn. When the weather changes, the leaves go from an ordered state (neatly attached to the tree), to a disordered state (scattered about the ground). Our intuition tells us that it is easy to imagine leaves falling from a tree down to the ground, but hard to imagine the leaves jumping up from the ground and reattaching themselves to the tree. This intuition is actually a reflection of the Second Law of Thermodynamics, which determines for us the natural direction of the leaves' motion: from tree to ground, and not the other way around.

The changes in entropy as described by the Second Law of Thermodynamics also reveal that heat flows from a higher to a lower temperature, a phenomenon that we experience frequently. Let's say a cup of hot herbal tea is brewed and the phone rings. That the tea cools during the phone conversation seems to be an obvious outcome. The heat from the tea is flowing to the room; as that happens, the tea cools off and the room gets just a bit hotter. Now imagine setting out a cup of room temperature tea and then walking away for a phone conversation. The idea that the cup would spontaneously heat itself up, becoming hot tea while you are away, seems preposterous. And in fact, it is preposterous.

The leaves falling to the ground and the hot tea cooling off are all different expressions of the Second Law of Thermodynamics at work. This law determines the direction of those processes. And, regrettably, it also requires that those processes will have losses that show up as waste streams. While the First Law of Thermodynamics indicates that the best you will ever be able to do with any particular energy system is break even, the Second Law of Thermodynamics says you will not even be able to do that. Those losses are the cause of the environmental impacts of energy, and those losses often impact the water system through thermal or chemical pollution.

The Second Law of Thermodynamics manifests itself in the form of inefficiencies or losses during energy conversion. Because of entropy, which increases for a closed system, inefficiencies are introduced, causing losses. These losses show up as waste heat, lost fuel, or suboptimal operation of systems. Inefficiencies are simultaneously a vexing problem and an enticing opportunity for the global energy system. In the United States, we consume about 100 quads (or quadrillion Btus) of energy each year. More than half of that energy—about 55 quads—is rejected as waste heat into the atmosphere from our smokestacks and tailpipes or into our waterways from cooling our power plants.[4] That means we have the confounding situation where we waste more energy than we actually use. If we could only find a way to successfully harness our waste streams, then we will have made significant progress toward solving our energy problems. Those waste streams include food waste, municipal solid waste, agricultural waste (such as manure), wastewater, and the waste products that come out of smokestacks in the flue gases, such as waste carbon dioxide and waste heat. The latter two are relatively abundant, and distributed wherever people and combustion take place, which happens to be a convenient location because it's next to where people often need energy.

No conversion from one form of energy to another occurs with 100 percent efficiency, so there are always losses for every conversion.[5] One way to think about efficiency is by measuring what you get out of a conversion compared with what you put into it. Highly efficient systems will convert more than 90 percent of the incoming energy into useful output energy, though in a different form. Some anthropogenic processes, for example electricity generators and boilers, are very efficient, while others such as steam turbines and incandescent lightbulbs are not. Surprisingly, many processes found in nature, while resilient and robust, are not very efficient. For example, photosynthesis typically has efficiency lower than 1 percent for converting energy in photons from the sun into chemical energy stored in a plant's biomaterial.

Unfortunately, the Second Law of Thermodynamics means that electricity production typically wastes about two thirds of the energy content of the original fuel. That is, for every one hundred units of fuel energy that enters a power plant at the start of the process, only thirty-three units of useful electricity leave the plant, and sixty-seven units of waste heat are generated. These effects are exacerbated by the losses from transmission and distribution, followed by the lightbulb itself. Including all the losses end-to-end, usually less than 1 percent of the original energy content of the coal is used to illuminate a room. For every one hundred units of fuel energy in the coal at the start, less than one unit of useful light energy is generated. That outcome is a travesty, and is one of the reasons why the Nobel Prize for Physics in 2014 was awarded to Isamu Akasaki, Hiroshi Amano, and Shuji Nakamura, inventors of an efficient blue light-emitting diode (LED). The invention opened up the pathway for white LED lights that are twenty times more efficient than incandescent lightbulbs.

Automobiles are also inefficient. The next time you get into your car, realize that only a small fraction of the energy contained in the original crude oil is being used to move you to your next location, with the rest being released into the atmosphere as heat and other waste products, including pollutants.

A related concept is that entropy can be reversed, but only by investing energy. That is we can reattach the fallen leaf to the tree, but to do so requires energy to pick the leaf up, climb the tree, and then affix it in place. We can clean our dorm rooms, but only if we invest effort. We can clean up our pollution, but it takes work to do so.

Another aspect of entropy and the Second Law of Thermodynamics is the conclusion that nonrenewable energy reserves are bound to become depleted. As we pull oil, coal, and gas out of the ground, the amount that remains is smaller than before. And if we keep up that process, eventually we will run out. Or a more likely scenario is that the prices to extract the fuels will increase to the point where they are uneconomical and other options will be more attractive. This point is in important contrast with water. The volume of water in the world is for all practical purposes fixed: it does not deplete and it does not grow, though it might move in place and time to be less available or dirtier. Thus, in this important way, energy and water are different.

All in all, the Second Law of Thermodynamics is a powerful and important edict. It says that our energy systems are and always will be wasteful. It says that we cannot avoid environmental impact from our energy conversions. And it says our conventional energy sources will eventually run out. If we heed its rules, then we would probably seek to design a system that is less wasteful, has smaller environmental impact, and uses fuels that replenish. Many of those options have water implications, too.

One of the most useful outcomes of the First Law of Thermodynamics is that energy can embody many different forms. To work our way through these forms, there are different classifications in the modern parlance we use for our convenience. One of the key distinctions is between primary and secondary energy.

The primary energy supply is made up of the original unconverted fuels in natural form, such as petroleum, natural gas, coal, biomass, flowing water, wind, solar, and uranium. Secondary energy is the converted fuels, including such forms as electricity, hydrogen, and stored energy. The difference is that you can "mine" for primary energy, but not for secondary energy. We can mine for coal and uranium and drill for oil and natural gas, but you cannot mine for electricity. Electricity, as a secondary form of energy, must be produced from something else. That is too bad, especially given how convenient, quiet, and clean electricity is to use with our appliances.

There are only a few different original sources for these primary fuels: the earth, moon, and sun. The earth provides radioactive materials for nuclear energy and geothermal resources that can be used for heating and cooling. However, much of the geothermal energy is generated from the heat of decaying radioactive materials, so that source could also be deemed nuclear. The moon is a source of energy, as its gravitational pull provides tidal forces that can be harnessed for mechanical or electrical power. Other than nuclear, geothermal, and tidal energy, all other energy forms originate from the sun.

Solar energy comes to us in a direct form, as incoming electromagnetic radiation—or light—that can be converted into electricity through photovoltaic (PV) panels or heat that can be turned into steam using mirrors that focus the solar beams. That steam can be used to spin turbines for making electricity or to drive a propulsion system for an old-fashioned Stanley Steamer automobile. Plus, there are many other forms of energy that are indirectly created by solar energy. For example, global wind patterns are caused by solar energy. Cycles of solar heating and cooling of the continents and oceans create the temperature differences that drive wind. That makes wind an indirect form of solar energy. Subsequently, waves are driven by wind, making wave energy a twice-removed form of solar energy. The sun also is a primary driver of the global hydrologic cycle, with evaporation and subsequent rain driven by sunshine. In fact, much of the sunshine that comes to earth is used to evaporate water from the oceans, doing the heavy lifting of raising water into the atmosphere for us.[6] So the energy forms embedded in flowing water—such as power derived from hydroelectric dams, natural thermal differences in the ocean between the relatively warm surface and cooler depths (ocean thermal energy conversion or OTEC), river currents, and salinity gradients where fresh and saltwater meet—all originate with the sun.

Notably, the sun provides the key input—sunshine—for photosynthesis, which allows plants to grow as feed, food, fuel, feedstock, and fiber. These items can be broadly categorized as bioenergy, and they represent solar energy that has been stored over time as chemical energy in the bonds of the plant's materials. Crops typically represent solar energy that has been stored for months, while old-growth forests represent solar energy that has been stored for decades or centuries. Solar energy has been the primary source of energy for most of mankind's historical endeavors, including farming, raising livestock, building houses with wood, making cloth from plants directly with cotton grown by the sun or indirectly with wool from sheep that were fed plants grown by the sun, and ensuring our survival by eating. It is only recently—since the 1860s—that mankind has engaged in large-scale use of an older, fossilized form of solar energy that was trapped in old plants and marine life that has been compressed and heated for hundreds of millions of years to form coal, oil, and natural gas.

These new fuels that gained significance in the 1860s are the fossil fuels. Although fossil fuels had been used for thousands of years—even the Bible refers to ointments, which might have been from natural oil seeps in the Middle East—that use was at a very small scale. It was only in the 1860s that their use grew to be a nontrivial fraction of the overall energy supply. Ironically, fossil fuels are simply a collection of old bioenergy that had stored solar energy over many millennia. This solar energy stored in biomass was then subjected to geological pressures and timescales that converted forests and swamps into coal and algae into petroleum and natural gas. Thus, even fossil fuels are a form of solar energy, though with a much slower rate of replacement, beyond the duration of human life, than the conventional notion of solar energy. Adding complexity, savvy nuclear scientists are quick to gloat that the sun is basically a massive nuclear power plant. Solar energy—and therefore all fossil fuels—are really derivative, indirect forms of nuclear energy.

Our familiarity with energy comes from the ways we buy and use it. We buy cords of firewood, gallons of gasoline, barrels of petroleum, tons of coal, cubic feet of natural gas, and kilowatt-hours of electricity. We use natural gas or wood to heat our homes and businesses. We use petroleum products to operate our machines in factories and our cars and trucks. And, we use many fuels to make electricity. Electricity is particularly valuable, because once you have flowing electrons you can use them to make heat and to move machinery.

In addition to using these resources as a fuel that is burned, their important role as feedstocks for manufacturing materials is often overlooked. Petroleum is used to make plastic, pesticides, pharmaceuticals, cosmetics, paints, dyes, and cleaners. Natural gas is used to make fertilizer, ink, glue, and paint, among other products. Wood is used to make paper, fenceposts, and other building materials. Coal is used as a source of heat and carbon for steel and iron production. Coal is also used in the cement-making process. And, the solid wastes from coal combustion, including bottom ash, which comes out the bottom of coal boilers, and fly ash, which flies through the smokestack, are used to make drywall for buildings and aggregate for roads.

Because there are so many forms of energy with different end-use applications and implications for society, they are organized into different categories. Labels such as fossil fuels, alternative energy, renewable energy, sustainable energy, green (or clean) energy, and unconventional energy color our conversations. Unfortunately, these different classifications are sloppy at best, and intentionally misleading at worst.

The (conventional) fossil fuels include coal, natural gas, and liquid petroleum. The key consideration regarding fossil fuels is that they were formed tens to hundreds of millions of years ago—in the fossil era—and are considered to be finite and nonreplenishing. According to the Second Law of Thermodynamics, if we keep using them, we will run out of them someday. However, just to make things confusing there are renewable forms of natural gas, sometimes called biogas or renewable natural gas (RNG), that can be produced from decomposing organic matter such as food waste or manure from agricultural operations. And, there are people who manufacture synthetic fuels (synfuels) such as synthetic gasoline and synthetic diesel, which are fuels with similar characteristics to the petroleum products, but made from coal or with some fraction of biogenic sources.

Unconventional fossil fuels include nonliquid forms of petroleum such as oil shale, shale oils, oil sands, tar sands, and heavy oils. Unconventional forms of natural gas include shale gas and coalbed methane. The word choices are interesting, as it is not obvious at first blush what the difference is between oil shale and shale oil. Oil shale is the kerogen that is essentially a form of rock found in Utah and Colorado. Shale oil, also known as tight oil, is the liquid produced from impermeable shales. In the former case, the oil is the shale. In the latter case, the oil is held by the shale and has to be cracked open by hydraulic fracturing or some other technique.

And, these words are wrapped up in different agendas. A college schoolmate of mine was a senior drilling engineer producing oil for one of the major oil companies in Alberta. In one of my communications I discussed the growing production of the tar sands in Canada, and he wrote back a terse email, excoriating me for referring to them as "tar sands" instead of "oil sands." To paraphrase his reply, tar sands conjure up images of strip mining operations with significant environmental impact, whereas oil sands are relatively cleaner and an abundant, secure resource. The words matter.

For a variety of reasons—their carbon emissions, the fact that they are depleting, and because some forms are imported—fossil fuels have become unpopular with large swaths of the population and with policymakers. And, they seem particularly unpopular in certain circles of the current generation of college students. In fact, it is not unusual for me to field inquiries from eager students who want to join my research group to do research on alternative energy. I usually begin by asking them "alternative to what?" In typical usage, alternative energy, renewable energy, and clean energy are often used interchangeably even though they mean different things. Usually what these prospective students broadly mean is "clean energy." But distinguishing between "alternative" and "clean" is important. Petroleum was an alternative to whale oil, and coal was an alternative to wood. Early in the atomic age, nuclear energy was considered an alternative to fossil fuels. However, today in the modern parlance, nuclear is typically one of the options for which an alternative is sought. And today, natural gas is an alternative to coal in the power sector, though both are fossil fuels. Also, the context becomes important depending on the end-use. For example, electricity is widely considered an alternative transportation fuel for cars, despite electricity being powered almost completely by coal, natural gas, and nuclear. And, natural gas as a fuel for vehicles is also considered an alternative to petroleum, even though natural gas is often produced alongside petroleum.

Renewable energy typically includes any form of energy that is renewed continually or annually. This category includes energy forms that are not depletable, such as solar, wind, water, and tidal. It also includes forms that renew quickly, but are depletable, such as bioenergy. "Renewable" refers to the characteristic of the fuel's availability and the rate at which it renews itself. Fossil fuels are not considered renewable because they replenish themselves over tens of millions of years, which is a timescale too slow to be practical from a human perspective. By contrast, no matter how much solar energy we use on a Tuesday, just as much solar radiation will hit the earth on the following Wednesday, so the solar radiation "renews" itself daily. Even though renewable energy comes back regularly, it is not inherently sustainable.

The dividing line between renewable energy and sustainable energy is not always clear. Roughly, these two are used synonymously, though they have distinctly different interpretations. In particular, it is possible to use renewable energy in unsustainable ways. It is possible to cut down trees faster than they grow back. Just because trees are renewable does not mean we use them sustainably. Forests typically take many decades or centuries to grow, but it only takes years or a few decades to cut them down, as was witnessed in the upper Midwest and northeastern United States in the 1800s.[7] So, while renewable energy is a classification dependent on whether a fuel is replenished quickly by nature, sustainable energy is a comment about the rate of consumption. To illustrate, solar energy is inherently sustainable: it is rapidly renewable—it comes back every day—and no matter how much we use, there is still sufficient solar energy. By contrast, finite forms of fossil fuels are inherently unsustainable: even if they are used at a slow pace, the replenishment rates are even slower and so the amount of geological resource in place only declines. Keep in mind that even though the resource will decline, our ability to extract it might increase with time, so the actual production rates might increase in the foreseeable future. Other renewable forms of energy also are not necessarily sustainable. While geothermal energy is renewable as it is driven by internal heating of the earth, some geothermal sites get played out. And hydroelectric dams can silt up after a century or so, limiting the sustainability of their usefulness despite the renewability of the water flows that power them. Dredging to remove silt extends the lifetime of dams, but that costs a lot of money and effort. Overall, it is important to remember that just because an energy form is renewable does not mean it is necessarily sustainable, clean, or green.

Green (or clean) energy are those forms of energy that have small environmental impacts. This idea is particularly contentious to define as it mixes technical characteristics of energy sources such as their emissions and behavioral choices about how we produce or use those energy sources. Clean energy once referred to whether it was clean at its point of use, and was primarily a reference to air quality. However, the vision has expanded to include the entire life cycle, such as extraction and production of the fuel itself (which implies a renewable or nondepletable resource base), processing (which implies a low energy intensity for upgrading and distributing), and at its use (which implies a low carbon intensity and toxicity). But all energy choices have an environmental impact: even clean options like wind energy have effects on land use, and solar energy has impacts from pollution at the mines that produce silicon used in manufacturing photovoltaic panels.

The world consumes a lot of energy from many sources for a variety of applications. Despite the diversity of options, fossil fuels—coal, petroleum, and natural gas—are the dominant historical primary energy sources and still provide approximately 85 percent of the world's energy today.[8] The rest of our energy comes from nuclear and renewables such as wind, solar, geothermal, bioenergy, and hydro. We put that energy to work, using those fuels in different ways for the different end-use sectors.

The transportation sector is highly dependent on petroleum, with more than 95 percent of the energy for moving our cars and trucks coming from gasoline and diesel. The reason for this reliance on petroleum-based fuels is that they are particularly good transportation fuels, with high energy density that makes them convenient to store on board the vehicle.[9] We essentially have a monopoly of fuels for transportation: we can choose our retail vendor—Exxon versus Chevron, for example—but we do not have much choice about the fuel. When people declare that we have an "oil problem," they are essentially saying we have a transportation problem.

In contrast to the transportation sector's dominance by a single fuel (petroleum), the electric power sector draws on a much more diverse array of fuels. While still heavily dependent on coal for more than a third of its energy, the power sector also draws heavily from nuclear, natural gas, and renewable sources. And, as coal prices increased steadily through the first

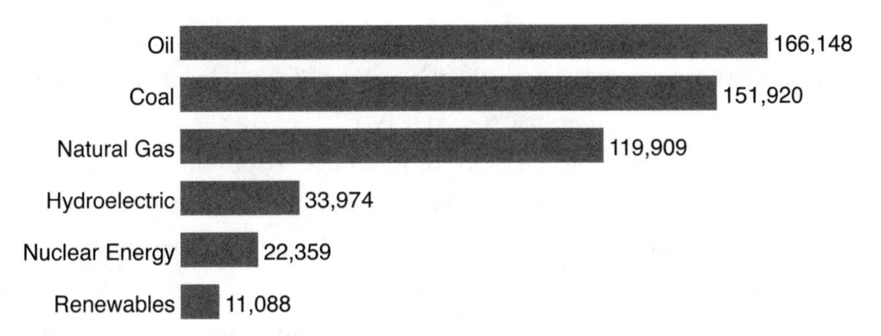

Annual energy consumption by fuel, globally, in trillion British thermal units, or Btu. [BP Statistical Review of Energy, 2014]

decade of the twenty-first century, and as natural gas and renewable prices leveled or dropped, that distribution has diversified further in the United States. Because it is so expensive, very little petroleum is used for power generation in countries where other options are available, serving mostly as a backup source with diesel generators or for peaking units during the hottest or coldest hours of the year when demand is highest. In contrast with the other forty-nine states, the power sector in Hawaii is heavily dependent on imported petroleum because there are no local sources of other traditional fuels and oil is the easiest to ship and store.

This point about the power sector's lack of dependence on oil is very important. Many observers in the United States are fond of saying we need more wind power as a way to reduce oil imports. Environmentalists and conservationists are fond of saying we should turn off our lights to reduce oil consumption. But these statements do not make any sense: we use very little oil in the power sector, so, outside of Hawaii, saving electricity does not save oil. We should still turn off our lights, but the way to reduce our oil consumption is through changes in our transportation sector or by replacing our dirty old fuel oil heaters in our homes with something better.

In addition to the different fuels and end-uses, the scale of global annual energy consumption is impressive. Overall, the world consumes more than 525 quadrillion Btu of energy.[10] One Btu is about the same energy as in a kitchen match, so that means the world's 7 billion people consume 525 million billion kitchen matches each year. Even more notably, despite having less than 5 percent of the world's population, the United States is responsible for nearly one-fifth (100 quads) of that global energy consumption. U.S. consumers use the same types of energy as global citizens: we just consume a whole lot more of it. Roughly speaking, the average global citizen consumes about 75 million Btu of energy each year. The average U.K. resident consumes twice as much as the average global citizen, and the average U.S. resident consumes twice as much as the average U.K. resident. So U.S. residents consume four times as much energy per person on average as the typical resident of China or India.

But, there are only 60 million British, and 315 million Americans. That leaves about 7 billion others who consume energy like global average citizens. If they all tomorrow started consuming energy like the British, then we would need to double annual energy production and consumption. If they consume like the Americans, the energy system would have to quadruple. And, the world's population is growing. By 2050 there might be 9 to 11 billion people. Factor in the additional people who also want to consume energy like Americans, and then it's hard to imagine that the earth's atmosphere or oceans could take it. It's also not clear that the global energy industry could meet the demand. Is it even possible to extract, refine, move, consume, and clean up that much energy at that rate?

This conundrum is essentially the grand challenge of the twenty-first century: how do we bring the value of that energy consumption—the clean water, indoor lighting, comfortable quality of life—to every global citizen in a way that does not leave behind a wake of environmental destruction?

While the modern-day energy policy debates hinge on a snapshot in time of current energy consumption and production, one of the most important lessons to learn is that the energy system is continually changing. Many of the key factors for energy vary with time, and there are many timescales for these changes. For example, we use decades or centuries to describe energy transitions from one dominant fuel to another, such as the transitions from wood to coal and coal to oil. We use years to decades to contemplate macro demographic trends such as population growth or for the construction of major energy facilities. We use months to years to evaluate seasonal shifts in energy production and consumption due to changes in the weather or other prevailing conditions that affect renewable energy sources' production or the demand for air-conditioning. We use hours to describe shifts in demand for electricity from baseload (at night) to peak load (in the afternoon on a hot day), and we use minutes or seconds to balance the power grid.

It turns out that preventing the failure of the power system requires mastering the second-by-second variations of electricity demand along with the multidecade process of grid planning. This mismatch in timescales is rife with problems and seemingly invites disaster at every step as some solutions for rapid grid-balancing such as sophisticated algorithms do not work for the political-economic-cultural process of building large-scale long-lived capital assets like power plants or oil export facilities that are used to produce or distribute conventional forms of energy. A capital lock-in effect of bad decisions ensues: if a company builds a $5 billion power plant that is expected to operate for forty years, they generally want to use it for the entire lifespan of the design so that they can get their money back. That means it is important for us to understand the broader trends before committing ourselves to expensive, big-ticket items that we might come to regret.

Over time, energy use has changed significantly. The dominant fuel from antiquity through the late 1800s was wood and other forms of bioenergy, which were used for heating, cooking, and as a feedstock for materials (such as lumber and paper). Starting at about the time of the Industrial Revolution in the mid-1800s, coal and petroleum production increased. Coal had already been used minimally in the early 1800s in a few discrete applications, but then was adopted more widely as a fuel for domestic and industrial use starting around 1850. In parallel, oil production started in earnest in 1859 in Titusville, Pennsylvania.

By 1885, coal had surpassed wood as a primary fuel in the United States. There were several reasons for this transition, including coal's superior characteristics as a fuel and deforestation throughout the upper Midwest and the northeastern United States. Many of today's beautiful forests in New England, Wisconsin, and upper Michigan are second-growth forests. By the late 1800s, logging had been so extensive that many of these

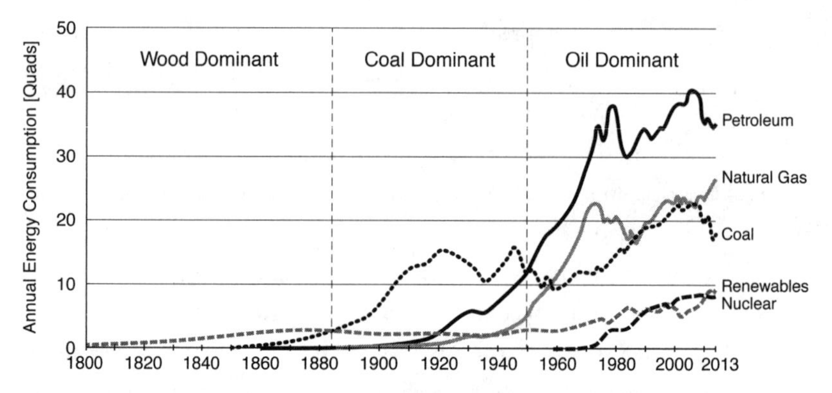

Annual energy consumption (in quadrillion Btu) in the United States from 1800 through 2013, showing transitions from one dominant fuel to another. [U.S. Department of Energy, Energy Information Administration]

forests had been almost completely wiped out, leaving an indelible environmental footprint and making wood more expensive and harder to come by.[11] Because of wood's increasing scarcity and price, alternative fuels were particularly appealing, which made coal—an affordable alternative to wood—an attractive option.

In addition, by most measures coal was simply a better fuel than wood. It has twice the energy density of wood at approximately 20 million Btu per ton of coal versus 10 million Btu per ton of wood, generates less smoke and ash, is less carbon intensive per unit of energy, and produces higher temperatures when burning. That last point is particularly valuable, as it made coal a better fuel for metalmaking applications. Prior to the use of coal, wood and char had been used for metalworking. However, because wood's combustion temperatures were lower, workers had to put more energy into the metal by hammering against an anvil, which explains the stereotypical image of the strong, sweating blacksmith hammering heated metal.

The unexpected irony here is that in the 1800s coal essentially saved the forests. Because a better and more abundant fuel—coal—came along, the rampant deforestation slowed down dramatically. This concept is difficult to digest in the modern context, for which coal is considered a threat to the forests through mountaintop removal mining, which wipes out entire mountains, valleys, streams, and forests, and through acid rain (composed of sulfuric and nitric acids), formed when coal combustion emits pollution that mixes with water vapor in the air.

Petroleum use grew at about the same time, but much more slowly. Initially, petroleum consumption was almost entirely in the form of kerosene as an illuminant and polyolefins as lubricants. Similar to the trajectory that coal faced, petroleum became popular because the conventional fuel sources—animal fats for lubricants and whale oil for lighting—were becoming scarcer and more expensive. While many people argue today whether peak oil is worthy of concern, in the mid-1800s, peak whale was a very real phenomenon: whale oil production peaked around 1850.[12] Literature from the time even featured whaling, revealing its importance to society. Herman Melville's *Moby-Dick,* published in 1851, was practically a technical manual for whaling.

These new fossil fuels had good performance. While the lubricants made from animal fats would clog up the new machinery that was becoming prevalent during the Industrial Revolution, oil-based lubricants had less friction and better consistency over a wider range of operating temperatures and pressures. Those improved the coal-fired machines' operation and helped spawn more growth in factories and automation. At the same time, kerosene was a much better illuminant than whale oil. It burned brighter, burned longer, generated less smoke, and did not have the pungent odor that was produced by burning whale oil. So, once again, depleting renewable resources (animal fats or whale oil) were displaced by the new fossil fuel alternative, which was more abundant, cheaper, and had better performance characteristics.

Just as coal saved the forests, petroleum saved the whales from extinction. Today, petroleum is considered by many to be a threat to whales and other marine life because of risks that are induced by noise and spills from coastal pipelines, offshore production, and ocean-borne tankers. These two examples are illustrative of another broader philosophy about energy: today's energy solutions often become tomorrow's energy problems.

Ultimately, petroleum use did not surpass coal use in the United States until approximately 1950. While petroleum was growing in popularity, the use of oil for lubricants and kerosene for lighting was small overall compared with coal and wood for heating and materials. However, with the advent of the internal combustion engine for automobiles, which transformed society near the start of the 1900s, petroleum found additional applications. After that, its use grew rapidly for several decades. For transportation, petroleum's high energy density and liquid form (which made for easy handling and portability) yielded much better performance in terms of power and endurance than biomass, which was the conventional fuel for other forms of transportation, namely trains, which used wood for boilers, and horses, which were fueled by feed. John D. Rockefeller, who had already become the world's wealthiest man from selling kerosene, became even richer once he could also sell gasoline.[13]

These historical transitions from wood to coal to oil reveal three important aspects: transitions take a long time; transitions have a distinct trend toward decarbonization; and there has been an unmistakable pattern of growth in energy consumption since the Industrial Revolution.

While the ability for a society to shift its fuel mix over time is evident, it is also clear that these transitions do not happen quickly. While biomass was the dominant energy source for millennia, its modern use as a source of energy for activities beyond sustenance and shelter picked up in earnest in the late 1700s to early 1800s with wood as its preferred form. Wood's decline as the dominant fuel source in the United States spanned centuries, until it was surpassed by coal in 1885. Subsequently, the rise and fall of coal as the most popular fuel source spanned sixty-five years, from 1885 to 1950. After that, petroleum has reigned supreme for more than sixty-five years, and continues today as the nation's most popular source of energy, though I expect natural gas to overtake petroleum in the United States by 2025.[14] If society's goal is to wean ourselves from fossil fuels entirely, then history's lesson is that it will take a while and so we better get started. Because these trends take so long, one role of government might be to set policies in place that accelerate the transition toward lower-carbon, sustainable fuels.

In addition to the long duration of these transitions, they also show a distinct trend of decarbonization. Wood emits more carbon per unit of energy content than coal. Similarly, coal emits more carbon than petroleum. Natural gas, whose primary molecular constituent is methane, CH_4, is the least carbon intensive of the four. As we switched from wood to coal to petroleum and then to natural gas, we have been preferentially selecting fuels with lower carbon intensity. Despite the decreasing amount of carbon emitted per unit of energy, the total carbon emissions in the United States increased throughout the twentieth century up until about 2008 because total energy consumption grew by so much. But our fuel choice became cleaner along the way.

This trend, which spans centuries, makes very clear that decarbonization is not a modern concept invented by Al Gore or environmental warriors. Rather, this pattern indicates that decarbonization is what societies do as they get richer. Civilizations naturally desire to "clean up their act" over time. In that context, modern efforts to combat global change through policy efforts that seek decarbonization would not be a departure from business-as-usual, as many critics allege, but rather would be a direct continuation of a trend that has been in place for more than one hundred years. Though to be clear, the prior shifts were not targeting carbon directly—they were targeting cost, performance, or cleanliness—and the decarbonization was a useful by-product. The difference for the ongoing transition is that carbon reductions are explicitly targeted because of climate change concerns.

Alongside our changing fuel mix was growth in total energy consumption since the early 1800s. This trend is the consequence of population growth—more people, each of whom consumes energy—and economic growth—more rich people, who consume more energy than poor people. In addition, society was undergoing several shifts, including urbanization, industrialization, electrification, and mobilization, each of which brought along with it additional energy requirements.

Economic growth typically implies higher per capita energy consumption as people gain affluence, and that trend is evident for many decades since the early 1800s. However, with many industrialized countries, per capita consumption leveled off or even dropped slightly since the energy crises of the 1970s, as the industrial mix shifted from highly energy-intensive industries such as manufacturing and chemical production to less energy-intensive service-oriented industries such as banking and research not to mention wide-ranging investments in energy efficiency as a way to reduce energy costs. In the early 2010s, as energy prices remained high and new appliance efficiency and fuel economy standards kicked in, our per capita energy use dropped further.

Increasing energy consumption brought with it many benefits, such as prosperity, better health, and improved quality of life. But that energy consumption also brought along several downsides. As energy consumption grew, so did energy imports, environmental degradation, emissions, and global climate change. For people who are energy poor, increases in modern energy consumption usually lead to a better life. Electric lights help poor students in remote villages do their homework at night, and clean-burning stoves spare their users from the risks of inhaling too much smoke and soot. But, for those who are already energy rich, increases in energy consumption actually might worsen our quality of life as the accumulating effects of the environmental impacts from energy consumption start to undermine the benefits. Solving this balance of good and bad is the main challenge moving into the future.

THREE

Water

If there is magic on the planet, it is contained in water.

—Loren Eiseley, *The Immense Journey*, 1957

WATER IS IMPORTANT TO LIFE, ecosystems, and most of the processes we care about. When NASA's deep space probes look for life, they look for water. Water helps give DNA its shape, which means water gives life its shape.[1] It is also a critical harbinger of environmental and ecosystem health. It has been said that the modern environmental movement in the United States launched in 1969 because of three separate, but important, water events: an oil spill, a burning river, and a trip to the moon. In January and February 1969, a massive spill from offshore production in the Santa Barbara Channel released over eighty thousand barrels of crude oil that lined the nearby beaches of Southern California and killed thousands of birds. Decades later, there is still oil and tar stuck in the sands, and a pipeline leak in May 2015 that released over two thousand barrels of oil along the Santa Barbara coast was a stark reminder that the risks to water have not gone away completely.

A few months after the 1969 spill, on June 22, the Cuyahoga River in Cleveland, Ohio, caught on fire because of its rampant pollution, capturing the attention of *Time* magazine.[2] Although water quality is not a subject most would consider themselves experts on, people do understand that water should not burn.

Nearly a month later, on July 20, the first manned mission successfully landed on the moon. Looking back at the earth from space revealed the globe to be more beautifully blue than people had ever expected. The space-borne view of the oceans served as a reminder that despite our planet's name, its surface is primarily covered by water.

Despite its complex role in life and the ecosystem, water's chemical makeup is quite simple: H_2O. Its chemical name has many variations, including dihydrogen monoxide, hydrogen hydroxide, hydric acid, hydroxic acid, hydroxyl acid, hydroxilic acid, and μ-oxido dihydrogen. Its structure is simple, symmetric, and nonlinear.

Water has a strong dipole moment, which makes it a polar molecule. In layman's terms, that means the charge is unequally distributed such that it "points" in a particular direction. While that concept might not have seemed significant during high school chemistry when we usually hear about it for the first time, polarity has several important consequences. For example, the polarity causes airborne water vapor to absorb a lot of infrared radiation, making it the atmosphere's most important naturally occurring greenhouse gas. In fact, water's contribution to atmospheric warming is why we have such a comfortable climate in the first place. Without so much water vapor and pre-industrial carbon dioxide in the air, and without its propensity to absorb radiation, our atmosphere would be about thirty-five degrees Celsius cooler; water vapor is responsible for approximately twenty of those degrees.[3]

Water's polarity also makes it a valuable universal solvent. It can dissolve all sorts of materials, including salts, sugars, and acids. And that's why we use it for cleaning. Water can also exist in three phases in nature simultaneously: solid icebergs floating on a liquid ocean below a sky with clouds of water vapor. Its strong dipole moment also makes water "sticky," which means it adheres to other surfaces as well as other water molecules. The net effect of this stickiness is that water has high surface tension, a feature that offers many benefits to living things that need to circulate water through their systems.

Water also has a relatively high heat capacity and heat of vaporization. These attributes mean that water can carry a lot of heat before it boils. The converse is also true: it takes a lot of energy to boil water, as anyone who has ever had to boil water on a campfire with sticks they collected themselves can attest. Because of its heat capacity, water makes an excellent coolant for power plants. For example, water's heat capacity is about four times greater than air, which means it would take four times as much air movement to accomplish the same level of cooling as water. If you burned your hand, you would probably prefer to cool it off with running water, rather than simply putting your hand in front of a fan.

One often overlooked unique feature that is very important is that water's maximum density is at four degrees Celsius. While the significance isn't obvious, it means that ice floats on water, a characteristic not shared by many other molecules. While floating ice is such an intuitive and obvious part of our observational experience, such behavior is unique. And critical. If ice did not float, but rather sank as most other solids would do, then lakes would fill up with ice.

Instead, the ice forms at the top layer, hovering above a large body of liquid teeming with life. If the density of ice were higher than liquid water, then as the ice was formed it would sink to the bottom of the water body. Each subsequent layer of ice formation would do the same thing, stacking up at the bottom of the water body until the entire lake was one giant ice block. As can be imagined, such a phenomenon would squeeze out the living creatures, depriving them of a chance to thrive. But, because of water's lower density, the cycle of freezing and thawing can proceed while life continues below, a convenient outcome.

The global hydrologic cycle is large, powerful, and continuous. It can change its intensity over time, but the cycle does not stop. And, different parts of the cycle are all interconnected. As described by the U.S. Geological Survey, "Earth's water is always in movement, and the natural water cycle, also known as the hydrologic cycle, describes the continuous movement of water on, above, and below the surface of the Earth. Water is always changing states between liquid, vapor, and ice, with these processes happening in the blink of an eye and over millions of years."[4]

The hydrologic cycle includes major fluxes and volumes of water. The largest movements of water are evaporation and precipitation over the ocean. There is also significant transport of water vapor in the atmosphere, rainfall and snowfall over land, and other fluxes in the form of runoff, stream flow, and evapotranspiration, which is the evaporation of water through photosynthetic activity from the growth of plants. It's clear that we have plenty of water: it is just that the water is in the wrong form (saltwater), place

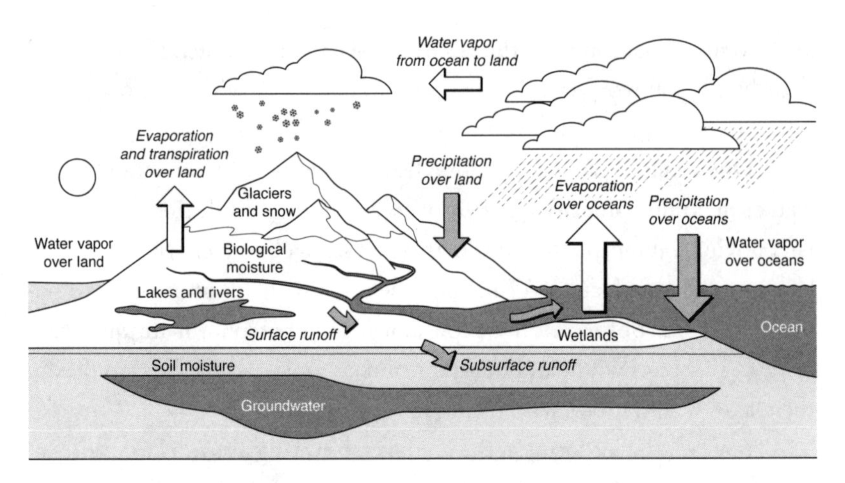

The hydrologic cycle is global and complicated, being composed of large storage systems and fluxes. Easily accessible surface freshwater makes up a small fraction of the world's water supply.

(on top of mountains, for example), or the wrong time of year. The image of the hydrologic cycle is essentially a depiction of water abundance, as long as we have energy to convert the water to our purposes.

The key forcing function that is the pump driving this cycle is the incoming solar energy from the sun. Just over half of the earth's incoming solar radiation is consumed in the process of evaporating water.[5] Evaporation of water over the ocean is the primary driver of the hydrologic cycle: once that stops, the whole cycle will come crashing to a halt. Essentially the sun does a lot of heavy lifting for us, raising water to a high altitude in the atmosphere, after which gravity brings it back down as snow and rain. If we could capture the entire gravitational potential of that elevated water, it would give us energy at a rate of 13 terawatts, which is nearly an order of magnitude higher than the rate at which the entire globe consumes electricity. As it rolls back down to the oceans, we harness it for power, irrigation, drinking, and many other purposes. Then the whole cycle starts again.

In other words, atmospheric water, despite constituting less than one-thousandth of a percent of the world's total water volume, is a key driver of the water cycle.[6] The good news is that there is a lot of water: the globe is awash in water. There is approximately eight times more water stored in

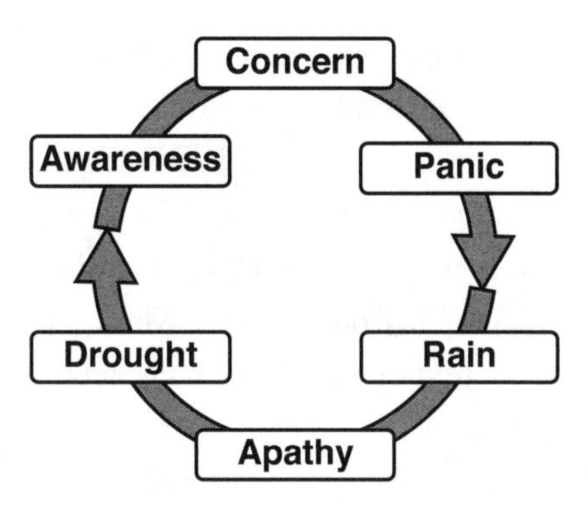

The hydro-illogical cycle shows the human reaction to the hydrologic cycle. Drought initially leads to water awareness, while prolonged drought raises concerns and ultimately panic, until rain causes people to be apathetic again.

the atmosphere than all of the world's rivers combined. There is 150 times more water in glaciers and snow than all of our lakes combined. There is plenty of water.

The old seaman's ditty "water, water everywhere, nor any drop to drink," from *The Rime of the Ancient Mariner* by Samuel Taylor Coleridge in 1798, captures the essence of inconvenience of being surrounded by seawater that is unsuitable for consumption. And, it is no surprise that the word "tantalize" has its roots in a water-based legend. The Greek gods punish Tantalus, a son of Zeus, by giving him great thirst and forcing him to stand in a pool of water that always recedes as he leans down to take a drink. Such a myth feels like a fitting parable for humankind's relationship with abundant water resources that seem to be forever just beyond our reach. In fact, it is this inconvenience that drives much of the energy investments for water: we spend significant sums of energy moving, treating, or storing water so that it is available in the form, location, and time we want it. While those energy investments overcome the limits of water's tantalizingly distant location, billions of people still remain without clean, accessible water.

While the hydrologic cycle is governed by the laws of physics, humankind's response is something entirely different. Tom Mason, the former general manager of the Lower Colorado River Authority, a major water and power provider in central Texas, introduced me to a new concept that sarcastic (and exasperated) water managers bill as the hydro-illogical cycle. This term came to life from planners who have decades of experience dealing with different stakeholders who need water. The hydro-illogical cycle is the series of steps that humans take as water goes through its various phases of availability. Just after rain, we develop apathy, after which drought catches our attention, leading to awareness campaigns, broad concern, then panic. When the rain comes again, we smile and start the cycle all over again. It is as if our entire public policy response to water challenges is to pray for rain and hope for the best. While that approach has worked many times in the past, sometimes it is not enough.

A critical underlying aspect to water science, which would be valuable to those frustrated planners managing the hydro-illogical cycle, is the importance of accurate water bookkeeping. To properly understand the state of water availability and the extent (if any) of the water crisis, it is critically important to track the flows, uses, and volumes of water in storage in different locations, forms, and times of year. Unfortunately, despite many relevant advances in the last century for hydrogeology, these are areas where scientists lag behind. The world's dataset on water is woefully incomplete, which makes it more difficult to make thoughtful water planning decisions. By contrast, energy data are relatively plentiful and available.

Keeping track of water is done through a scientific bookkeeping method known as the Reynolds Transport Theorem, which is used to track the flows, fluxes, and storage of water in a physical system. It's similar to the way we do our financial bookkeeping: the amount of money we have in the bank tomorrow is the sum of what we had in the bank yesterday, plus whatever we deposited today, minus whatever we withdrew. Whether we get rich depends on how much we start with and whether we withdraw more than we deposit.

The same is true with water: the key aspect is the rate at which water is withdrawn or consumed, and whether it exceeds the rate at which water is added to the system from rainfall or aquifer recharge. If we take water out of an aquifer faster than water comes back in, then our water bank account will go dry. If we withdraw water at a slower rate than our deposits, then the water bank account will fill up.

Those who have watched the children's animated movie *Rango* will recognize this concept. In that movie, the main character is a lizard voiced by Johnny Depp who finds himself stuck in a desert town. That town is going through a water crisis amidst a backdrop of political corruption, not unlike the storyline of *Chinatown,* the 1970s movie about water in Los Angeles that starred Jack Nicholson. In *Rango,* the town's bank account is literally a large jug of water. As the town's members had been withdrawing from the bank faster than the rate of deposits, the bank account—their store of water—was going dry. This movie is an illustration of a worldwide phenomenon as societies on every continent are withdrawing water faster than they are depositing, which means their accounts will eventually run out.

Unfortunately, in contrast with bank accounts—where we get a nicely printed record of our deposits, withdrawals, and balances—for the water world, we are operating in an information vacuum. While withdrawals might be known in some instances—for example by large industrial, municipal, or agricultural users who monitor their pumping—generally we do not know the status of our water resources or water use. Even in the United States, where governmental record keepers have a plethora of data on all sorts of matters, our data on water are not sufficiently updated, comprehensive, or specific to location and time.

Precipitation is one of the methods for water additions, and so important that governments monitor it closely. It can also be easily tracked, but how much of that rainfall recharges the underlying aquifers is hard to measure or estimate. Sometimes we do not know our water balances until the well runs dry. Such a situation is akin to writing checks for the things you need, such as the mortgage, groceries, and gasoline, without ever knowing for sure how much money you had originally in the bank, how much money remains, or how much money you are earning. Basically, the status of your account would be unknown up until you run out of money and the checks start bouncing. That is a reasonable approximation for our water situation today. We are operating blind while writing a bunch of checks that are likely to go bad someday.

Installing stream gauges to track flows in rivers, well monitors to estimate the water tables, and pump meters to keep track of withdrawals across watersheds would be helpful. This approach is implemented in some richer parts of the world. But we still do not know how much water is in the ground. However, new technologies such as GRACE give us new eyes in the sky that can see through the ground to tell us a little more about our bank account of groundwater.

The Gravity Recovery and Climate Experiment (GRACE) consists of twin satellites that were launched in 2002.[7] The way GRACE works is that the two satellites follow each other in their orbit around the poles of the earth. As they orbit, differences in the earth's mass cause slight differences in the gravity field, which causes the satellites to speed up or slow down in their orbit. While orbiting, the satellites send a microwave beam back and forth and they measure how long it takes the beam to travel between them roundtrip. Doing so lets them measure their distance from each other, which reveals tiny differences in their relative acceleration. Since acceleration depends on the earth's mass directly below, the oscillating distance between the satellites provides a highly detailed map of the earth's gravity field. As they map the earth's gravity over the years, the satellites paint a detailed view of how the earth's mass changes with space and time.

The main point here is that the earth's mass is not uniform the way people might think. For example, some newer mountain ranges (the Andes and Himalayas come to mind) have more mass than their low-lying neighbors. Because water is so heavy, the mass of different regions—such as Brazil—will vary based on whether it is the wet season or the dry season. That means GRACE can be used to measure changes in the mass of water from space. Amazingly enough, satellites hundreds of miles high in the sky can see through the ground to tell us whether our aquifers are filling or going dry with greater precision than our farmers working on the surface.

Unfortunately, the early results have been alarming: the aquifers in India are emptier and depleting faster than scientists thought, and the ice sheets in Greenland and Antarctica are melting faster than anticipated.[8] GRACE revealed that in northern India, the water table dropped about one foot each year between 2002 and 2008 as farmers overpumped the aquifer to irrigate their crops. It is nice to have that information for the first time about the amount of water in our bank account. Whether that information will cause us to be more frugal with water remains to be seen.

Just as there are many different forms of energy, there are also different forms or types of water. These types differentiate the water's composition, where it came from, and the different labels we assign them.

One of the most important distinctions has to do with the salinity or the level of total dissolved solids, or TDS (made up of salts and minerals), in water. Potable water has less than 1,000 milligrams per liter of total dissolved solids.[9] "Brackish water" is too salty to drink (with a TDS of 1,500–10,000 milligrams/liter). Above that is "saline water" (10,000–100,000 milligrams/liter), which includes seawater. The wastewater that is produced out of the ground from hydraulic fracturing ("fracking") of shale formations can have a TDS as high as 400,000 milligrams per liter, which is one of the reasons fracking is so contentious.[10]

In Spanish, freshwater is called sweetwater (*de agua dulce*), as its flavor is much sweeter than that of saltwater. If water has a salt content that is too high, then drinking it will actually do physiological damage by poisoning the body's cells. At the same time, perfectly pure water with no dissolved solids (such as distilled water) can also cause damage by leaching salts and minerals back out of the body's cells. This is an unfortunate consequence, especially since there are vendors who sell distilled water as a "healthy alternative" to conventional water. Those salts and minerals also provide taste. While a relatively high level of salts tastes bad, it turns out we do have a preference for a small dose of salts that provide flavor. In fact,

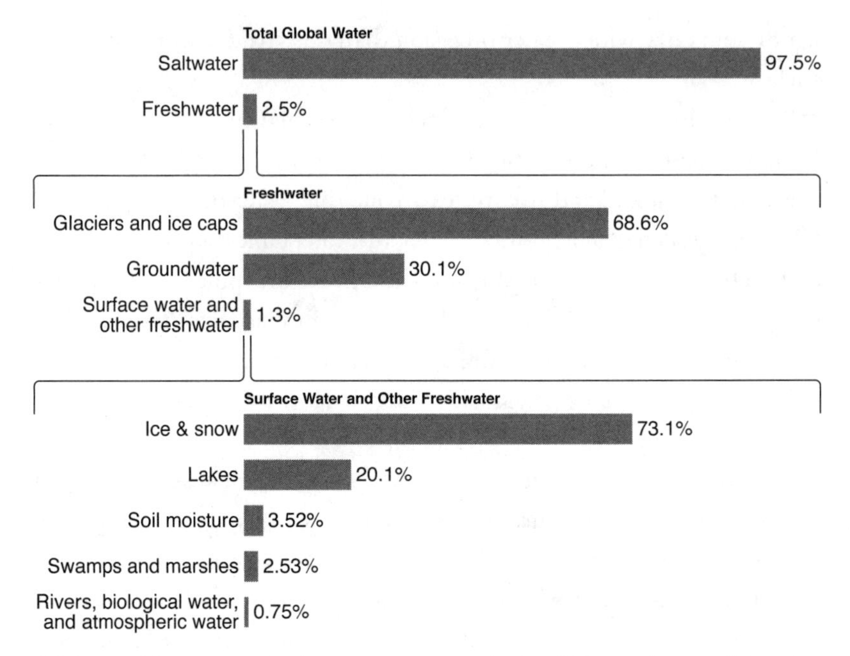

Of the world's water, only about 2.5 percent is fresh. Of that, only a small fraction is easily accessible surface water in rivers and lakes. [Peter H. Gleick, editor, *Water in Crisis: A Guide to the World's Fresh Water Resources* (New York: Oxford University Press, 1993)]

many bottled water purveyors actually add a prescribed mixture of salts to achieve a particular taste.[11]

Rivers have a range of salinity, depending on where their water comes from, the composition of the riverbed, pollution to which they are exposed, and so forth. Generally speaking, rivers are potable, though there are some very polluted rivers that cannot be drunk because their level of salts, toxics, and contaminants exceed acceptable thresholds.

For human consumption and agricultural production, freshwater is necessary. However, there are a few crops that grow with brackish water, and there are sea plants such as seaweed, kelp, and algae that grow in saline water. While the world is awash in brackish and saline water, the relative abundance of freshwater is really what matters for human prosperity and quality of life. Unfortunately, freshwater is a small fraction of the total:

about 97.5 percent of the world's water is saline or brackish (1.365 billion cubic kilometers), and only 2.5 percent (35 million cubic kilometers) is freshwater.[12] Of the world's freshwater, much of it is locked up in ice or other locations that are hard to access.

In addition to the salinity of water, it is also useful to track the location and source of liquid water. The first distinction is between surface water and groundwater. Surface water sits on top of the ground in rivers and lakes (usually freshwater) or in the oceans (saltwater), ultimately covering 70 percent of the earth's surface. Groundwater is water in the ground (as opposed to on the ground) and can be fresh or saline. An aquifer is a geological zone with a large volume of water that is naturally stored in porous rock such as sandstone.

Of the world's freshwater, more than two-thirds is in glaciers and permanent snow cover, about 30 percent is in groundwater (including soil moisture, swamp water, permafrost, and aquifers), and only a small fraction of freshwater is in lakes and river storage.[13] That means there is a hundred times more water in the ground than in all the world's rivers and lakes. In other words, very little of the world's water is easily accessible surface freshwater.

Water in the soil above the water table is called the unsaturated zone, and that section has relatively less water by volume. So if you plan to dig a well, you have to dig it deep enough so that it reaches the saturated zone, where water is more abundant. And, as groundwater sources become depleted, the water table drops. The saturated zone gets lower, and wells have to be dug deeper. The deeper you have to dig, the more energy you have to spend to pump the water to the surface.

While for planning and legal purposes, surface water and groundwater are sometimes treated as separate and unique, they are in fact connected. Surface water often trickles down into the ground, recharging aquifers. And, there are places where the groundwater comes out naturally or under its own force. Those are called springs if the outlet is natural and flowing artesian wells if the outlet was manmade. Artesian wells are dug, drilled, or cut deep into the ground at a section of the water table that is under enough pressure such that the water comes to the surface on its own.

Prolific wells can produce more than three hundred gallons per minute. Such wells conveniently allow us to avoid the nuisance of having to lower a bucket or use a hand pump to raise the water. Sometimes the pressure is so high that the water shoots out of the well several feet. Such an arrangement—with the earth doing the pumping for us—is handy. And it's also safe, as it removes the risk of careless people or unattended children falling down a conventional open well as in *Tikki Tikki Tembo,* the children's story set in ancient China. It's not just a children's story, either. Toddler Jessica McClure fell down a well in her aunt's backyard in 1987, creating an international media circus as rescuers worked around the clock for more than fifty hours on live broadcast television to free her. They were successful, and she is alive and well today.

In addition to groundwater versus surface water, it is also useful to distinguish between renewable and nonrenewable water. The hydrologic cycle describes the fluxes of water around the globe. Water evaporation is followed by precipitation is followed by aquifer recharge or runoff, then evaporation again. Consequently, water is considered renewable in that it renews itself: it will rain again someday.

However, the rate of renewal is important to consider. For aquifers, water is added to the porous rock through a process called recharge. Aquifers all have a different rate of recharge. Pumping water out of wells faster than the rate of recharge will eventually deplete the aquifer—causing your well and your neighbors' wells to run dry—despite the fact that water is renewable.

Some aquifers recharge and discharge very quickly. By contrast, portions of the Ogallala Aquifer, which spans several states, have what some people bill as fossil water: the water is millions of years old in some places and has a very slow recharge rate. This recharge is so slow, in fact, the water stored in the aquifer is essentially nonrenewable: it is similar to fossil fuels, which are stored energy underground.

Just as energy has different colors—green energy and brown coal come to mind—we can also think about water's different colors. In James McBride's autobiography, he recounts his story as one of twelve children of a black father and a white mother in an era where interracial couples and mixed-race children were considered outcasts.[14] His tale includes the passage to adulthood along with struggles for race identity, the pursuit of success, and overcoming (or redefining) racist stereotypes. In the story's defining moment, James asks his mother if she's white. She replies that she is "light-skinned." He asks her whether God is black or white, and she answers, "God is the color of water." In other words, since water has no color, God is colorless. God is neither black nor white.

While I suspect it is true that God does not have a skin color, and water is nominally colorless, it is also standard to use colors for describing different types of water, and from a variety of perspectives. There is of course the classic whitewater, meaning quickly flowing water in rivers, in which the stirring makes it appear white. But in addition, there are also different color schemes for water footprinting, and for use in buildings, natural environments, and national security.

The Water Footprint Network provides the definitions to differentiate blue, green, and graywater.[15] Blue water includes consumption from fresh surface and groundwater (lakes, rivers, aquifers) from a water footprinting perspective. Green water is precipitation on land that does not run off or recharge the aquifer (stays in the soil), but might eventually be evaporated or evapotranspired during growth of a crop. Graywater is water that becomes polluted during production or is needed to dilute pollutants.

In the built environment graywater is the polluted wastewater that is generated as residue from washing and looks cloudy or gray from the soaps and other organic matter. It includes the water from sinks, laundry washers, showers, and tubs. It can also include the stormwater harvested, for example, by collecting runoff from the roof during rainstorms through gutters that lead to rain barrels. But it does not include sewage from toilets. Blackwater is wastewater that contains sewage (fecal matter and urine) and organic matter from dishwater drains. It is stored in a septic tank or transported away to a treatment facility where pathogens are removed.

Some advanced eco-sensitive homes separate the graywater and blackwater, using the former on-site for irrigation and sending the latter to a facility for treatment. While that does reduce the need for treated freshwater, this approach can be problematic for sewage collection systems if the flows get reduced so much that they do not work properly. Purple water is treated wastewater effluent that does not meet potability standards, but is suitable for applications such as irrigation and washing cars. Many municipalities paint the pipes purple to provide a visual clue that the effluent is not potable.

Taking the analogy of water and financial bookkeeping a little further, rainfall is similar to our income: like the money added to the bank from our salaries, rainfall is the water that is added to the local system. The difference is that rainfall is free, given to us by a combination of natural factors, namely the hydrologic cycle, which elevates water to the atmosphere to the benefit of all of us, laboring as a hard-working water pump that works for nothing and without complaint. Rainfall makes water renewable—it brings the water back to us after we have consumed it and sent it down the rivers to the ocean. Rain recharges our rivers, lakes, and aquifers. Because it is one of the easiest parts of the water cycle to see and measure and because rain is like our income for the water bank account, it is closely tracked by companies, citizens, and governments.

Because rainfall is so central to the health of the U.S. agricultural sector, the U.S. Department of Agriculture is one of the key agencies that supports this effort. Rainfall varies from less than 4 to greater than 160 inches per year for the contiguous United States. Texas is a microcosm of the national range, varying from less than 14 inches in the arid west near El Paso and exceeding 54 inches annually in the piney woods and swamplands near Houston. It is surprising to many people that Houston, in dry Texas, has average annual precipitation that exceeds precipitation in Seattle (38 inches) in the rainforests of the Pacific Northwest. The main difference is that the precipitation in Seattle occurs rather steadily throughout the year, whereas Houston's rain comes in a few great storms. In fact, the Texas water cycle is often described as ravaging decades-long epic droughts punctuated by biblical floods. A typical joke about the weather in El Paso goes like this: "El Paso only gets 14 inches of rain each year, and you don't want to be there the day it happens."

While it is important to know how much rain falls and where it falls, there is more to the story: namely, *when* matters, too. Whether the rainfall is distributed evenly throughout the year or comes in bursts like the monsoons in southern Asia drives the rhythms of agriculture and design of the local societies. In Asia, whose monsoon season governs much of the culture, some places suffer drought for eleven months, only to endure torrential rains and floods from the monsoons for the remaining month.

Simply stated, floods are a combination of extreme precipitation and changed ecosystems that cannot accommodate the extra water. The rate at which the rain falls is very important. At a slow rate, the rain can accumulate into lakes and reservoirs, trickle down into aquifers to recharge them, or soak into the soils, grasses and trees. A healthy ecosystem works to absorb the water. However, if the rate of rainfall exceeds the rate at which it can be absorbed, or if the ecosystem has been impacted in a way that reduces its ability to absorb water—for example, if parking lots have replaced fields or woods, or if paved canals have replaced natural rivers—then a flood is a likelier outcome than in prior eras. Floods are one part nature, and one part humanity. And the human part is a bigger contributor to floods than people realize. By building homes in floodplains, paving over grassy meadows, and bulldozing wetlands, we have diminished nature's ability to absorb the water when it surges in. And we get floods that have higher propensity to kill, destroy cities, and wipe out infrastructure.

Droughts are important, too. Droughts are expensive.[16] They can ruin crops, kill livestock herds, and inhibit recreation such as boating, floating, or fishing. The drought in California in 2014 caused losses in the agricultural sector exceeding $2 billion and 17,000 jobs. The one-year drought in Texas in 2011 led to $7 billion of damage. The one-year drought and heat wave in the midwestern United States in 2012 did $30 billion of damage and killed more than one hundred people. At its peak, the drought covered a majority of the land area of the United States and caused more than half of all counties to be disaster areas. Drought is a big enough concern that the U.S. government tracks it and makes the data publicly available through the U.S. Drought Monitor.[17]

All these factors—rainfall, drought, floods—combined with unequal population densities mean that renewable water availability per person is not uniform globally.

Historically speaking, it is unusual and highly valuable that the United States invests in collecting such high-quality rainfall data and makes them available. Imagine how beneficial such information might have been for ancient societies such as the Anasazi Indians of the desert Southwest or the Mayans, which struggled with persistent drought. While they might have also had very sophisticated means of record-keeping, it is unlikely they possessed a mechanism for distributing that information as broadly as we can achieve today with Internet-enabled communications. Nor did they have satellites available to check the amount of water remaining in underground bank accounts.

That we have detailed data and monitoring available is a testament to our ingenuity as a society. But it also means that we cannot claim ignorance for our decisions that make our water situation worse. We might be operating blind in many ways, but we are much better equipped with scientific data than ancient societies. And so if the wells run dry, we will only have ourselves and our shortsighted thinking to blame.

Jill Boberg, a researcher at the RAND Corporation, wrote an excellent report entitled *Liquid Assets* that identifies the demographic factors around the world affecting water use: population, urbanization, standards of living, prevailing mix of economic sectors, and number of households.[18] As these factors change—the population grows, the number of households increases, and societies switch from agrarian to industrial modes—water demands will shift, too.

In particular, the number of households and household size—the number of residents, not the square footage of the house—seem to be key drivers of consumption. As noted by Boberg, "Per capita, smaller households consume more water and produce more waste." And, while the number of households is increasing globally, the number of people living in each household is decreasing. As we get richer, multigenerational households are shifting: instead of cramming three generations under one roof, where we share the laundry, cooking, and outdoor watering—we are setting up individual households, driving up water use.

The per capita water use in the United States is relatively high. Because the United States is relatively affluent, the per capita withdrawals are quite high for municipal use for drinking, washing, and irrigating our lush lawns. In addition to being rich, the United States is a major agricultural producer (in contrast with other rich countries, such as Japan and the United Kingdom). Those agricultural needs drive water withdrawals up even further. Then, the United States also has an extensive industrial sector, with significant power generation, chemicals production, refining, mining, and other water-intensive activities. Consequently, North American water withdrawals (including Canada) are much higher than in other regions of the world. That might be fine for Canada, which possibly has more freshwater than any other country in the world, but for the United States, it can be a strain.

These withdrawals can also be contemplated in terms of per capita water availability, which varies globally. In this context, availability means the amount of water that is accessible to people over the course of a year within a reasonable distance. While Asia has the most total water available, it also has the highest population, and subsequently the lowest per capita water availability. This measure of water's abundance and availability to users is important when evaluating how trends in water withdrawals, coupled with the use of nonrenewable water sources, will trigger significant water strains.

Water use in the United States is significant: every day we withdraw about 400 billion gallons and consume 100 billion gallons.[19] That works out to more than 1,300 gallons per person per day withdrawn, of which more than 300 gallons is consumed per person per day. Withdrawals refer to the water that is taken out of a water body, some of which is consumed and some of which is returned. Consumption refers to the water that is evaporated or otherwise lost; instead of being returned to the water source, it soaks into the soil or comes down as rain somewhere else.

That water is withdrawn from a mixture of sources—groundwater and surface water—and from a variety of source qualities, including saline and fresh, for a mix of end-uses. While the U.S. daily withdrawals for the power sector are higher than for agriculture, globally, agriculture is a much larger cause of withdrawals. That's because the United States and other richer societies use much more electricity, whereas poorer societies are much more agricultural. And, similar to the situation with energy, residents of the United States withdraw about twice as much water per person as Europeans and four times as much as Southeast Asians. Europe gets more rainfall, making irrigation less necessary, and Europeans have more favorable attitudes toward conservation. Also, while the power sector is responsible for the greatest volume of water withdrawals in the United States, the agricultural sector has the greatest water consumption. That consumption is from evapotranspiration during photosynthesis, when water moves up and out of plants; runoff; and water that trickles back down to aquifers.

Water also serves an important role for transportation. It's well known that railroads are more energy efficient at moving goods than trucks, cars, or planes, but many people do not realize that waterborne commerce is even more energy efficient—as long as you have the water in the first place and don't mind the risk of spills and water contamination. In the United States, there are 11,000 miles of inland waterways that move half a million ton-miles of freight annually, which is nearly a third of the 1.7 million ton-miles of freight moved by 141,000 miles of railroads.[20] In particular, the lower Mississippi and its famous barges move a lot of cargo. In addition, almost all intercontinental freight (a result of globalization) is moved by water. This is especially true for the vast amount of crude petroleum that travels by supertankers on the oceans and barges moving coal and refined goods through canals and inland rivers.

Water projects are a hallmark of a civilized society. They are also closely tied with politics, as water projects are often important symbols of political power built by and named for politicians. In the nineteenth century, Abraham Lincoln ran on a platform of enhanced water infrastructure—namely, canals—for navigation and commerce.[21] Decades later, the largest dam in the world was built and eventually named for President Herbert Hoover. The first hydroelectric dams were not very large and were more reminiscent of the medieval overshot waterwheels that were used for mechanical power to grind grain, cut wood, or polish glass. The flowing water of the river rotated a massive wooden wheel that was connected by wooden gears and axles to a workhouse on the bank of the river. Early dam builders might create a small reservoir ten feet high, but generally speaking, these structures were not considered to have that great an impact on the river's natural flow.

After waterwheels had been used for hundreds of years, the first dams for generating electricity were built in the 1800s. The first one in the United States was at Niagara Falls in 1882. It simply diverted some of the natural flowing force of the water above the falls, and did not need to build a reservoir. One of those early dams, in Austin, Texas, spanned the Colorado River (not the same Colorado River that carved out the Grand Canyon in Arizona) and was called "The Great Dam." When it was built in 1892, it was the largest dam in the world and was featured on the cover of *Scientific American.*

The Great Dam was built for the same reasons as other early dams: for power, flood control, and irrigation. It also collapsed twice, over the first few decades of its life, foretelling one of the major risks of dams. One of those collapses had the fortunate outcome of creating an island in the middle of the lake in downtown Austin that today serves as a dog park where I occasionally take my dogs for a swim, so sometimes these unfortunate circumstances can yield a positive outcome.

Today a dam of the same size stands in the very same spot, and it has a powerhouse with 16 megawatts of generating capacity. That capacity is absolutely tiny by the standards of a modern dam. That this little dam was once the world's largest is a stark reminder of how much dams have grown over time: from a few megawatts in power a century ago to the gargantuan Three Gorges Dam in China today, which is 22,000 megawatts in power. If the little Austin dam can fail, what happens when the big ones fail?

In the United States, the build-out of hydroelectric dams accelerated right after the Great Depression. Their pace and scale grew in tandem. The hydroelectric infrastructure was also tied to the economic recovery efforts during the Great Depression and to the military effort of World War II. The

Great Depression created the economic motivation for large public works projects, while the war created the demand for electricity. In the 1930s, jobs were scarce, and large water projects were seen as a way to keep people working while achieving other useful benefits such as providing power and meeting the needs for navigation, commerce, recreation, flood control, and water storage. This is when the Hoover Dam (initially named the Boulder Dam) and several other prominent dams—the Shasta, the Grand Coulee, and so forth—were built. Dams today are still justified based on the multifaceted benefits they offer. For example, the Sardar Sarovar Dam in India was designed to provide irrigation for a million farmers, drinking water for 29 million people, 1.5 gigawatts of power, and jobs for five thousand employees.[22]

Though the dams offered a variety of attractive benefits, they weren't without their problems. And those problems inspired some resistance. To help overcome that resistance, the U.S. government invested in propaganda to persuade unconvinced stakeholders. They made films to serve as commercials for how important those dam projects were. They printed posters. And they even commissioned the folk musician Woody Guthrie, who wrote "This Land Is Your Land, This Land Is My Land," to pen a song about the dams along the Columbia River. It worked: the dams got built.

Most of the early build-out in the United States was in the Pacific Northwest along the Columbia River basin and in the southeastern United States. The abundant electricity provided by these dams kicked off a large military effort located right next to the massive power plants. Because of World War II, which relied on airplanes more than any prior war in history, there was significant new demand for aluminum. Since aluminum is produced electrolytically from bauxite (by contrast, steel is produced thermally from iron ore), many aluminum smelters were located near the dams.[23] Abundant electricity enabled aluminum production at a pace that had never been seen before.

In addition, there was significant demand for enriched uranium for nuclear weapons. Since uranium is enriched with electrically driven centrifuges, the appetite for power was enormous. At one point during the peak of the war effort, 10 percent or more of national electricity consumption was dedicated just to enriching uranium.[24] Dams were a key piece of that effort, and consequently the main nuclear labs for uranium processing were established in Washington state near the Columbia River dams and in Tennessee near the dams built by the Tennessee Valley Authority (TVA). It is telling that some of the Department of Energy's main nuclear sites in the United States are still found in those locations: Pacific Northwest National Laboratory in Washington and Oak Ridge National Laboratory in Tennessee.

Dams are our modern-day temples. They are massive structures that it seems like politicians build in their own honor. And, according to Marc Reisner in *Cadillac Desert,* many of the dams in the United States were built partly as a consequence of bureaucratic rivalry between the Army Corps of Engineers and the Bureau of Reclamation.[25] Traditionally, the Bureau of Reclamation builds water projects for irrigation purposes, to reclaim the land, so to speak. The Army Corps of Engineers has a mandate that includes building, widening, deepening, straightening, and managing canals, levees, and dams for the purposes of flood control and navigation. Neither explicitly has a mission to build power plants, but the hydroelectric capacity of the dams they were building for other purposes would often provide the economic justification for the projects in the first place.

And so a bureaucratic race ensued to see who could build the biggest dams the fastest. Along the way, though, some negative impacts from large-scale water projects emerged: population displacement and ecosystem damage. Dams and the reservoirs they create can be incredibly large, and requires vast tracts of land. To gain the land for the projects, they had to flood many acres of ecosystem and displace a lot of people. Politicians do not build dams in rich areas: they build them in poor areas, displacing poor people. There are two main reasons for this phenomenon. First, the poor people lack the political power to prevent the loss of their land. Second, because dams require so much area, they are built where the land is cheap, which is also where poor people tend to live.

Because of the controversy inherent to this story, American movies captured this pattern. The movie *Wild River* in 1960, starring Montgomery Clift and directed by famed moviemaker Elia Kazan, addressed this tension. The Depression-era plot focuses on a stubborn matriarch who refuses to let her island in the Tennessee River get flooded by the Tennessee Valley Authority. It was the conflict between letting rivers run wild, versus the drumbeat of progress for electricity and flood control. *O Brother Where Art Thou,* set in the 1930s and starring George Clooney, also captures this theme about how "the South is gonna change" because of a new dam.

Another movie with this premise is *Deliverance* from 1972, which starred Burt Reynolds. The story follows four men who take a canoeing trip one last time before a river and its poor inhabitants are submerged and displaced by a new dam that is needed to "push a little more power into Atlanta for those smug little housewives." That means the poor rural people and the land were going to be displaced and flooded forever just to make some rich people in the faraway city a little more comfortable. The movie's theme—and opening dialogue—is one of rape: man rapes nature in the form of the dam cut into the hillsides. Then nature—in the form of a local hillbilly—rapes man.

As Reisner pointed out, there was great irony in what happened when dams were built in the West and the East.[26] When dams were built in the eastern half of the United States, they flooded fertile farmland in the name of controlling floods and generating power. At the same time, dams were being built in the western half of the United States as a source for irrigating the desert. That is, we were building dams to create farmland in the West to make up for the farmland being lost to dams in the East. Not only that, but the farmers who lost their land, along with the rest of the taxpayers, subsidized this ordeal. This whole thing might seem preposterous, but that was the path the United States selected for building major dams.

The ecosystem damage is hard to assess. While dams make electricity relatively cleanly, as there are no emissions at the point of power generation, they distort the landscape with a permanent mark. Fish and other animals cannot freely travel the rivers once dams are put in place. The old joke goes like this: "What does the fish say when he bumps into a wall? Dam." Famously, salmon go upstream to spawn, fighting currents to go uphill many hundreds of miles and jumping over obstacles along the way.

While they can impressively jump over small rapids a few feet tall, they cannot jump over dams. Consequently, some dams have installed "fish ladders" that are a cascading series of waterfall steps that allow the salmon to bypass the dam. But even if the salmon successfully navigate the fish ladder, their navigational systems, which benefit from a distinct current in the water, can get confused by the slack water on the other side of the dam. Fish can get caught going downstream, too. The whirling blades of the hydroelectric turbines have been known to filet the fish while they pass through. More recently there has been the advent of "fish-friendly" turbines that have broader spacing between the blades, which allows the fish to pass through with less likelihood of being injured.

There are the other ecosystem impacts, too. Silting is a major problem. Sometimes silt is needed downstream because its minerals are useful for agriculture. But dams stop the flow of the silt, causing it to accumulate behind the concrete walls. Eventually, the silt will fill up the reservoir, causing the dam to lose its function. This phenomenon happens for natural dams and lakes, too. Mirror Lake in the Yosemite Valley was a beautiful glacial lake that reflected some of that national park's greatest sites. It was formed by a natural dam of rocks and rubble left behind by the glaciers that caused water to pool up behind it. But over millennia, the natural lake has been filling up with silt such that it is now more of a pond than a lake and is filled only in the spring when the water levels are high. Eventually it will disappear altogether.

Because there is no combustion at the point of use, no greenhouse gas emissions are released during power generation at hydroelectric dams. However, greenhouse gases are released due to the energy consumed during construction of the dam. Furthermore, the large reservoirs release significant amounts of methane from the anaerobic decomposition of organic biomatter that was flooded during the filling of the reservoir.[27] Since methane is a very active greenhouse gas, these emissions are important to determine. Unfortunately, estimating how much gas bubbles out of the reservoirs is very difficult and rife with uncertainty.

There is also the undesirable impact that dams have on temperatures.[28] Usually temperatures are relatively uniform for a free-flowing river. But after a reservoir is built, the temperature can vary significantly from the water surface (relatively warm) to the bottom of the water column (relatively cold). When the water flowing through the turbines comes from the middle or bottom part of the reservoir, the water exits at a lower temperature than the temperatures to which the native river species are adapted. In some cold environments with surface-release dams, the opposite can happen: dams release water downstream from the relatively warmer surface, making downstream temperatures slightly higher than they would have been otherwise. In both situations, fish reproduction can be affected. Consequently, native river species must often migrate upstream of the dam to reach normal conditions or move downstream until temperatures stabilize.

Because of all these ecosystem and human impacts, resistance to large dams grew along with the dams themselves. By the late 1970s, dam construction had mostly halted in the United States. Most of the good sites were already taken, anyway, and the opposition to dams had become significant and well organized. For the most part, large dams are hard to build anywhere in the rich, developed world. Yet in developing countries, the allure of dams and all the economic, political, and electrical power they represent is hard to resist.

As a result, large dam construction is mostly taking place in poorer countries with growing economies. Those countries will get the same benefits—cheap electricity, flood control, navigation, and so on—but they will get the same problems, too. Flooded ecosystems, distorted fish migration, silting, and displacement of people are unavoidable. The Sardar Sarovar Dam in India displaced 150,000 people, and the world's largest dam—the Three Gorges Dam in China—displaced 1.3 million people.[29]

The Three Gorges Dam had been desired by Chinese leaders for decades, and it was finally constructed in the 2000s. While it has helped to reduce the risks of flood-related disasters and improved the navigability of the Yangtze River, its creation flooded entire valleys and towns, displacing people and erasing towns.[30] Geologists worry about the earthquakes and underwater landslides that the water causes as the soft, soaked soils around the reservoir settle to accommodate the new load. In the first decade of its operation, the reservoir, which is as long as Great Britain, triggered more than five hundred earthquakes with a magnitude greater than 2.0 on the Richter scale and more than four hundred landslides. If the Three Gorges Dam were to collapse, it would put approximately 15 million downstream lives or more at risk. In the event of a collapse, the biggest manmade wall of water ever conceived would move quickly down the canyons, making it difficult for people to escape. Unfortunately, more than six hundred dams are either built, under construction, or in planning in the seismically active Himalayas, putting the dams at serious risk of failure. If the Tehri Dam in India collapses, scientists expect it would produce a wall of water two hundred meters high that would put 2 million people at risk. While we can hope such a catastrophe will not happen, unfortunately, dams collapse every once in a while. And, when they do, the results can be horrific.

The Johnstown Flood of 1869, triggered by a collapsed dam, killed more than twenty-two hundred people in Pennsylvania. The dam collapse in Santa Clarita in 1928 killed six hundred people and is still reverberating through Southern California water politics. Even the movie *Chinatown* in 1974, starring Jack Nicholson and set in the 1930s, makes reference to the open wound of a collapsed dam. The collapse of the Grand Teton Dam along the Teton River in eastern Idaho in 1975 was caught on videotape, and it shows the powerful force of the water. Thankfully, it happened in the middle of the day, so only eleven people were killed. But the flood wiped out two small towns; if it had occurred in the middle of the night, thousands of people could have perished.

The scale and risk of the Three Gorges Dam, in terms of both its water and its power, are enormous. It is the ultimate testament to human hubris, evidence that we believe we can tame the earth to suit our wishes. One of the towns flooded by this dam is the City of Ghosts, which was built more than eighteen hundred years ago and contains temples and shrines dedicated to the underworld. In the novel *World War Z,* which is about a global zombie pandemic, the flooded City of Ghosts is also the source of Patient Zero. It is implied that our hubris and disrespect for the underworld spawned the zombies. Subsequently, when a child in the story was swimming in the reservoir behind the dam, he was bitten by one of those zombies, starting the outbreak. So dams save us from floods and drought and give us electricity, but they also put us at risk of man-made floods and, if fictional books are to believed, might cause a zombie outbreak.

In addition to the external water infrastructure—dams, canals, and reservoirs—there is also an extensive piped water infrastructure. That piped system includes the components for collecting, treating, and distributing drinking water and collecting, treating, and releasing treated wastewater. And it is vast, expensive, and leaky.

The build-out of the piped water system occurred in waves corresponding to population booms and economic booms that accelerated the expansion, punctuated by wars and economic downturns that slowed down the expansion.[31] Unfortunately, these piping systems fall apart over time. Pipe lifetimes have great variability, and depend on factors such as their use, local climate, pipe materials, and soil composition.[32] Shifting soils and tree roots put pipes under strain that can trigger their failure. The flow rates, variability of flow rates, water composition (including potentially corrosive components), and frequency of use all affect the stresses and strains on the pipes. Climatic conditions also affect the pipe's structure. Climates that cycle through freezes and thaws, or wet seasons and dry seasons, cause the pipes themselves and their environment to expand and contract or shift dramatically, putting additional strain on the pipes. The pipe materials themselves are also important. Cast iron pipes last longer than steel pipes, which last longer than PVC pipes. Also, soil composition is important: if the soils contain corrosive acids, they will eat away at the pipes, accelerating their decline.

Deterioration of the piping system is inevitable, so the U.S. Environmental Protection Agency conducted a "gap analysis" to determine the needs of the U.S. water infrastructure.[33] While every pipe will break eventually, it's hard to predict which pipe will break when. To approximate the risks, EPA scientists used an aging model with deterioration distribution for each pipe material. Pipes installed before 1910 with cast iron are assumed to last 120 years. Pipes installed before 1945 with steel are assumed to last 100 years. Pipes installed after 1945 are assumed to last 75 years. Ironically, their deterioration rates could coincide after 2015, which suggests that our water system might start crumbling faster than we are accustomed to.

That also explains why the U.S. water system is so leaky. Anywhere from 10 to 40 percent of municipal water is lost between the water treatment plant and our water meters at home. It could be also that the meters themselves are slowing down with time, so that some of the water is just unbilled rather than lost. But, that's not a good sign either. Unfortunately, fixing that crumbling, leaky infrastructure is likely to be very expensive.

Going back to the EPA's gap analysis, in addition to identifying infrastructure needs, the agency also estimated that roughly one-half trillion dollars of new investment would be needed to maintain U.S. drinking water systems at the quantity and quality that we want.[34] The state of Texas conducted its own projections and analysis for its State Water Plan, and identified a need for $53 billion of new investments just to guarantee sufficient water supply, even after accounting for significant conservation and efficiency.[35] That money is for reservoirs and other large-scale systems for catching, storing, moving, and treating water. Texas has roughly 10 percent of the U.S. population and economic activity, so it's no surprise that its estimate is roughly 10 percent of the national estimate. Dividing those water costs per capita looks expensive: approximately $1,500–2,000 per person. But keep in mind, those are the incremental costs on top of what we already pay for maintaining and operating our water systems today based on investments made decades ago.

And those systems are interconnected with energy, for better and for worse.

FOUR

Water for Energy

WATER IS A CRITICAL INPUT for energy. Overall, about 15 percent of the water the world currently uses goes to making energy in one form or another.[1] While most people are aware that water is used directly for generating power in hydroelectric turbines at dams, it is often overlooked how important water is for other parts of the energy sector. For example, water is used in the extractive industries for producing fuels such as coal, uranium, oil, and gas. In addition, water is an input for energy crops such as corn for ethanol or biomass for fuel pellets. Water is also a critical ingredient for the power sector as a coolant that increases the efficiency of power plants. Generally speaking, water improves the energy sector, which is great news. It also sets the energy sector up for vulnerability in the event that water is not available the way planners expect.

Globally, hydroelectric power is the largest source of nonthermal generation in use (the other common options being solar panels and wind turbines), accounting for over 16 percent of generation.[2] Thermal generation (also known as thermoelectric generation), which uses heat to boil water into steam or to pressurize other gases to spin a turbine, makes up almost all of the rest. Hydroelectric is a powerful, efficient, and reliable source of energy—as long as it has water. The water use implications of hydroelectric power differ significantly from thermal generation because it does not withdraw or consume water for cooling. Instead, hydroelectric facilities use the force of gravity to pass water through turbines to generate electricity.

The typical design is pretty straightforward: a dam is built to create a large reservoir of water with a significant elevation differential. The ele-

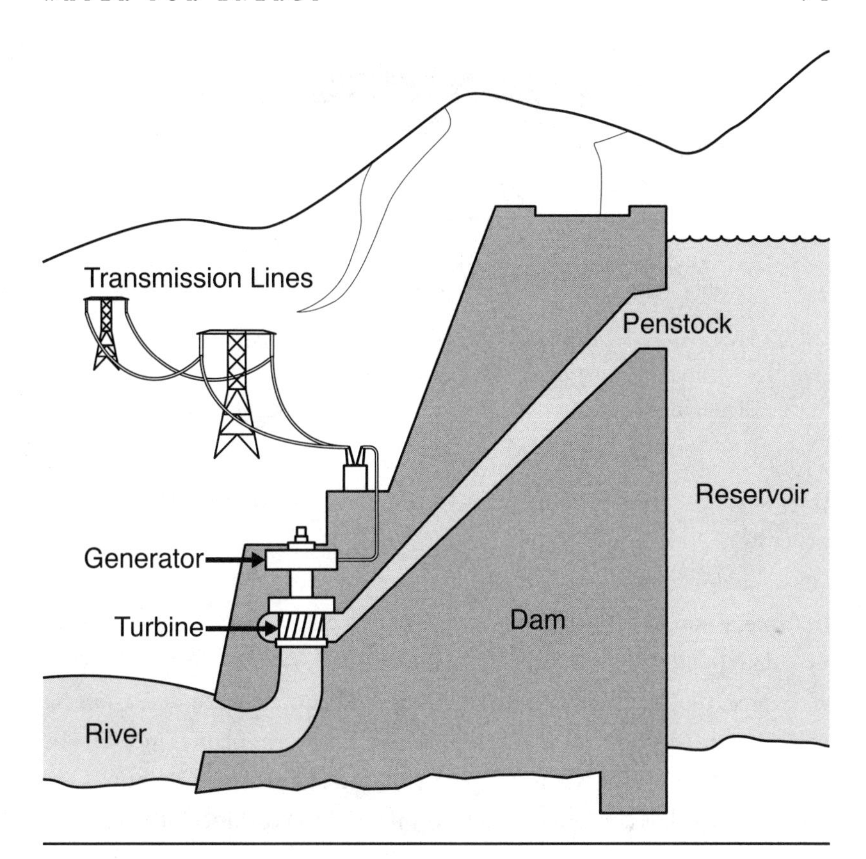

Water can be used to generate electricity. In a hydroelectric facility, falling water spins turbines that are connected to generators. Usually a dam is built to create a reservoir of water at a higher elevation than the river below.

vation difference between the water behind the dam and the river downstream of the dam creates potential energy that can be converted to mechanical energy (*m*) from rotating turbines that can be converted to electrical energy (*e*) from the spinning magnets within a generator.

The power output of the system is a product of the height difference through which the water falls and the volumetric flow rate of the water. Large volumes of water falling long distances at a high rate of speed generate a lot of power. The hydroelectric turbines rotate on a vertical axis, like a merry-go-round. And they are large devices. Water falls through

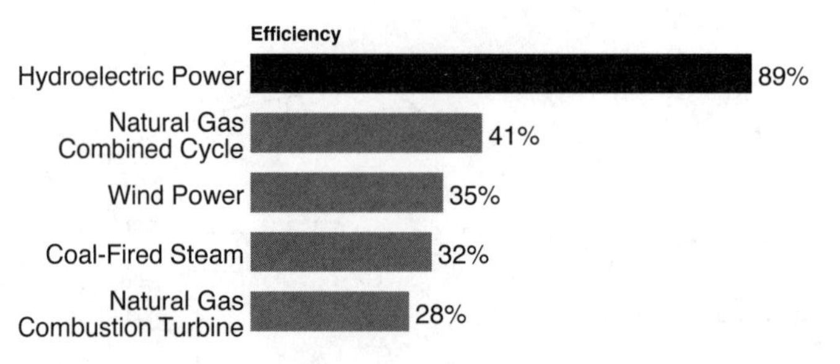

Hydroelectric turbines are very efficient, especially compared with conventional power generation options. [U.S. Army Corps of Engineers: Institute for Water Resources, *Hydropower: Value to the Nation,* Fall 2001]

the curved blades, forcing them to rotate a shaft that is attached to a generator.

The process is very simple, and consequently dams are highly efficient: they can usually achieve nearly 90 percent or higher conversion efficiency from the potential energy of the elevated water to electrical energy at the powerhouse.[3] This performance is much better than the 30–40 percent efficiency that is typical for conventional thermal power plants, and still one and a half times better than the most efficient, state-of-the-art natural gas power plants that combine steam turbines and gas turbines in a sophisticated fashion to achieve 60 percent efficiency. One of the key advantages noted before is that solar energy does the heavy lifting for us, raising water to high elevations from evaporation, after which we can harness the free force of gravity on the way back down.

Hydroelectric power plants can be absolutely massive, both in area and power generation. The largest power plant in the world, the Three Gorges Dam in China, has a capacity of 22 gigawatts, about the size of twenty or more nuclear power plants. It is so large, the mass of the water in the reservoir slowed the earth's rotation. By putting nearly 40 billion tons of water elevated to hundreds of meters above sea level, the dam has essentially made the earth a little fatter in the middle and flatter at the top, extending the day by six-hundredths of a microsecond.[4] The gargantuan Hoover Dam, which is famous as one of the world's first major dams and for its proximity to Las Vegas, is only 2 gigawatts by comparison. If the water supply is reliable, large dams can be used for baseload power. But hydroelectric power plants also have the ability to be quickly turned on and off, which gives them great operational flexibility. That means they can also be used to meet peak load or to firm up the power grid.

Because the construction of large dams has such a large impact on ecosystems, building new ones is contentious in most developed countries. While efforts to build major dams are still under way in Asia and South America, increasing hydropower generation in the United States and Europe faces a lot more resistance. Getting public approval for new sites can be quite difficult.

Consequently, for the United States, smaller-scale opportunities, known as "small hydro" or "microhydropower," or addition of hydropower at existing facilities that otherwise do not generate electricity have a better likelihood to be built.[5] The U.S. Department of Energy's Oak Ridge National Laboratory in 2012 determined that 12 gigawatts of new power generating capacity can be installed by adding powerhouses to some of the eighty thousand dams in the United States that are used for storing water, controlling floods, and managing navigable waterways, but do not make electricity. The existing fleet of twenty-five hundred hydroelectric dams has a total capacity of 78 gigawatts, so simply adding turbines to these nonpowered dams could expand hydropower capacity by 15 percent or more. Just one hundred dams of those eighty thousand could be used to add 8 gigawatts of capacity, which is similar to eight nuclear power plants. Because these nonpowered dams have already incurred many of the construction costs and ecosystem impacts, adding power to the existing dam site is often a faster, less expensive, and less controversial way to expand hydropower than creating new reservoirs that displace people and flood ecosystems.

Because smaller dams can be just a few feet tall, reminiscent of beaver dams, they are less disruptive. They can be built deep in the woods without flooding valleys or blocking fish migration, and generally keep their impact to a minimum. Smaller dams face less opposition and permits are easier to obtain. There are also run-of-river designs with the turbines laid along the riverbed without a dam at all, instead just harnessing the currents.

Experiments with these designs have been conducted in the East River in New York City. Enclosed versions can be inserted into downhill pipes like those in the California aqueduct. As water flows downhill inside the pipes, turbines spin, generating electricity. These versions are termed hydrokinetic, as they take the kinetic energy out of the flowing water to make electricity.

Then there are the ocean-based systems. The headline-grabbing ones include wave power and tidal power. Waves are continuous, driven by the wind with a force that beachgoers would easily recognize as powerful and unstoppable. Designs have emerged over the years to capture this energy. They include space-age-looking devices such as buoys with pistons inside that float up and down, gates that rock back and forth underwater on the ocean floor, snakelike contraptions that twist back and forth with the waves, and turbines-in-tubes at the ocean's edge that spin back and forth as water rushes in and out. The ultimate resource is large, globally, but expensive to harness as the waves themselves beat up the equipment and the saltwater is corrosive. Ultimately, turning wave power into a massive power source will include lining our coasts with hundreds of miles of these power plants, which is likely to be implausible given concerns about the possible effects on the marine environment.

Tidal energy is also appealing. Driven by the moon's gravitational pull, which conveniently raises the oceans several feet twice daily, this renewable resource does not deplete. And the technology is the same as for conventional hydropower, so this resource is reliable and simple. But the falling water is available only at certain times of the day and it requires good elevation gain between low tide and high tide. Those conditions happen in just a few places around the world—Nova Scotia and off the northern coast of France are two famous sites. That means the potential for tidal energy is not quite as universal as one might anticipate.

Other ocean-borne designs include Ocean Thermal Energy Conversion (OTEC) and salinity gradients. OTEC devices have a design a century old, exploiting the temperature difference between the relatively warm ocean surface and the cold ocean depths to drive a power plant. Differences in salinity at the mouth of rivers where freshwater mixes with the ocean can be used with osmotic membranes to create a flow of water to generate electricity. While that sounds promising, early Norwegian experiments were only able to generate enough power for a lightbulb. Undeterred, in late 2014 the Dutch announced a trial of similar technology with the hope of making it financially viable by 2020.[6]

Although hydropower does not require water for cooling like thermal generation, it is often considered a highly water consumptive technology due to the large volumes of water evaporated from the surface of reservoirs behind dams that house turbines.[7] The increased surface area of man-made reservoirs beyond the nominal run-of-river accelerates the evaporation rates from river basins. Notably, the estimates for this increased evaporation depend significantly on regional location. Major dams in the desert Southwest of the United States are especially prone to water loss from evaporation. Dams in cooler, wetter climates do not lose water as quickly (or at all). Furthermore, whether all the evaporation should be attributed to power generation is not clear, as reservoirs serve multiple purposes, including water storage, flood control, and recreation.

In addition to direct power generation through hydroelectric turbines, water indirectly enables power generation through the cooling it provides for power plants that use heat to make steam that drives a steam turbine. Overall, about 75 percent of the world's power plants use heat to make power. And those hot power plants need cooling to protect equipment and to make them more efficient. That amounts to a lot of cooling water. In exchange, the thermal power plants heat the water up before it is returned to the source.

Overall, the power sector is the single largest user of water in the United States, responsible for nearly half of all water withdrawals (a little more than 160 billion gallons per day, when including seawater), withdrawing even more than agriculture.[8] When considering only freshwater withdrawals, the power plants and agriculture withdraw about the same amount, roughly 115 billion gallons per day. Although agriculture uses significant volumes of groundwater and surface water for irrigation, the power sector primarily uses surface water. The mining sector, which includes the extractive industries for fuels production, requires another 5 billion gallons per day, and the industrial sector, which includes refineries and other facilities for upgrading fuels, is responsible for another 16 billion gallons per day of withdrawals.

Despite the power sector withdrawing the most water, the agricultural sector consumes the most. A vast preponderance of the water that is withdrawn for power plants is returned to the source, though at a different temperature and quality. The amount of water that is withdrawn and consumed by thermal power plants is driven primarily by a mix of factors including the fuel (coal, gas, nuclear, etc.), turbine design, cooling technology, and local weather. Nuclear power plants are particularly water intensive because, unlike power plants fueled by coal or natural gas, they cannot shed any waste heat to the atmosphere through smokestacks. Nuclear power plants don't have any emissions, so all of their waste heat has to be dumped into waterways.

The three most common cooling methods are open-loop, closed-loop, and air-cooling. On average across the thermal power sector in the United States, about fifteen gallons of water are withdrawn and just under one gallon consumed for every kilowatt-hour of electricity that is generated. Because typical homes in the United States use about twenty to forty kilowatt-hours of electricity each day, approximately three to six hundred gallons of cooling water are required to make electricity for those homes. That same home might use 150 gallons per day for washing, cooking, drinking, and watering lawns. That means we use two to four times more water at home for our lights and outlets than our faucets and showerheads.

Open-loop or once-through cooling withdraws large volumes of fresh and saline surface water, passes it through the power plant for one-time use, and returns nearly all the water to the source with little of the water being consumed due to evaporation. While open-loop cooling is energy efficient and low in infrastructure and operational costs, it has water impacts. The water intake systems at power plants can entrain and impinge aquatic life, also doing damage. Entrainment is the withdrawal of fish and aquatic organisms from the environment into the power plant facility. Impingement is the pinning of fish and aquatic organisms against water intake screens.

In addition, the discharged water is warmer than ambient water, causing thermal pollution, which can kill fish and harm aquatic ecosystems. As a result, environmental agencies regulate discharge temperatures, taking into account a water body's heat dissipation capacity. If power plant operators return the water above their approved temperature, they could be fined.

Closed-loop cooling towers are familiar as large, concrete inverse parabolas with white clouds of water vapor escaping out the top. People often associate them with nuclear power plants, but they work with other plants, too. Cooling towers withdraw water then recirculate the water until it evaporates, which has a cooling effect. Because the cooling is essentially achieved through evaporation, closed-loop cooling causes higher water consumption.

The need for such large amounts of water at the right temperature for cooling introduces vulnerabilities for the power plants. If a severe drought or heat wave reduces the availability of water or restricts its effectiveness for cooling due to thermal pollution limits, the fact that the power plant consumes so little water becomes less important than the fact that it needs the water in the first place.

Power plants constructed before the 1970s almost exclusively used open-loop cooling designs, which have very high water withdrawals. When these power plants were built, water was perceived as abundant, and environmental regulations were practically nonexistent. During the 1960s and 1970s, environmental concerns about water increased, kicking off an era of regulatory pressure to reduce water use at power plants. One of the key pieces of legislation was the Clean Water Act (1972), which established the framework for regulating discharges of chemical and thermal pollution into the waters of the United States.[9]

The Clean Water Act outlawed the unpermitted discharge of any pollutant from a point source into navigable waters. Point sources—discrete locations such as pipes or man-made ditches—are regulated by the Clean Water Act, but broader sources of pollution such as runoff over a wide area including farms and other agricultural operations are not. While homes do not generally need a permit for their wastewater flows into the sewers or septic systems, industrial, municipal, and other facilities must obtain permits for their discharges that go to surface waterways. In this way, the Clean Water Act regulates discharges from power plants. It also regulates intake requirements. Power plants built since then have almost exclusively used closed-loop designs with cooling towers as a way to serve many environmental interests by greatly reducing the entrainment and impingement of aquatic wildlife and reducing thermal pollution by limiting hot water returns.

Open-loop cooling withdraws more water, but consumes less. Closed-loop cooling towers withdraw less, but consume more. Which of these two options is "better" can depend on local prevailing circumstances. However, the conventional wisdom implies that cooling towers have less of an impact than open-loop cooling systems.

Today, 43 percent of U.S. thermal power plants are large power facilities with generation capacity of over 100 megawatts. Of these large power plants, 42 percent use closed-loop cooling towers and just over 14 percent use cooling reservoirs. Just under 1 percent use dry cooling, which is also known as air cooling, and the remaining 43 percent of these large power plants use once-through cooling.[10] Most of those plants with once-through cooling systems were built before the Clean Water Act was enacted or were grandfathered in once the legislation was passed. Many of these plants, built before strict emissions controls, are decades old and are simultaneously dirty and thirsty. Whether they should be shut down in exchange for newer, cleaner, leaner plants remains a hotly contested public policy debate.

Moving forward, new hybrid and dry systems might see greater implementation because of looming regulatory requirements and competition for water. For example, the California State Lands Commission proposed a moratorium on construction of new power plants with open-loop cooling systems on the coast. That clashes with a separate effort to push power plants to coastal regions where open-loop cooling can use seawater to spare inland freshwater.[11] In other words, environmental concerns about oceanic wildlife are in direct conflict with environmental concerns about inland freshwater supply. Conservation, efficiency, and the use of alternative, water-efficient options can meet those competing environmental objectives simultaneously.

More water-efficient cooling technologies exist; however, these systems have drawbacks. Dry-cooled systems operate like radiators in automobiles: a coolant is circulated within a series of closed pipes and air blows over it to cool the pipes. Air-cooled systems withdraw and consume less than 10 percent of the water of wet-cooled systems.[12] However, dry-cooling systems have higher capital costs and reduce overall efficiency of the plant, which increases costs and emissions per unit of electricity generated. Because the heat capacity of air is so much lower than water, much more air has to be moved to achieve the same cooling as with water. That means much larger facilities are needed to create the larger cooling surfaces in dry-cooling systems, which dramatically increases capital costs. Furthermore, a power plant with dry cooling can experience a 1 percent loss in efficiency for each degree increase of temperature, limiting power generation when it is hot outside.

Because they include both closed-loop wet systems and dry-cooling equipment, hybrid wet-dry cooling systems provide a compromise between wet- and dry-cooling systems.[13] Hybrid wet-dry cooling systems can have low water consumption for much of the year by operating primarily in dry mode, but have the flexibility to operate more efficiently in wet mode during the hottest times of the year when the extra cooling gives a boost in the power output from the plant. Unfortunately, water resources are typically less available during these peak demand times. Although dry- and hybrid cooling systems are proven technologies and useful solutions in water-strained areas, they usually are not economically competitive designs if water is cheap and available. And, in severely water-constrained regions, dry cooling is often the only alternative. In such cases, the up-front capital costs and reduction in the power plant's efficiency are more readily justifiable.

In addition to the water needs for hydroelectric power and for cooling conventional thermoelectric power plants fueled by coal, natural gas, and nuclear, the other forms of renewable power—solar, wind, geothermal, and biomass—also need water for their operation. The range of water they need varies dramatically.

Renewable electricity technologies such as wind turbines and solar photovoltaic (PV) panels do not use heat to make electricity, so they do not need cooling water. They need small volumes of water for manufacturing components at the steel mill where the turbine parts are fabricated or the semiconductor factory where the solar panels are printed, and they also use water for cleaning equipment in the field. Other than that the water needs are minimal.

The heat-based forms of renewable power—such as concentrating solar power (CSP) that uses mirrors to focus solar beams onto a central location to boil water, enhanced geothermal systems that use the heat of the earth, and biomass-powered plants that burn wood chips—all need water for cooling. The challenge with solar thermal plant systems is that areas that provide the best sunshine for CSP are typically dry and hot—think southern Spain and the American desert Southwest. As Roger Duncan, the former general manager of Austin Energy, is fond of saying, "What a solar thermal power plant needs to be effective is a desert with a lot of water." The conundrum of this situation is challenging. Although dry-cooling technology can be coupled, doing so introduces efficiency losses, particularly on hot days. Nonetheless, some solar companies have committed to dry cooling to avoid the political, availability, and environmental barriers posed by citizen concerns over water issues. These new systems demonstrate the feasibility of dry-cooling technology for large-scale systems and might be indicators of a new trend in electricity.

Geothermal power plants utilize naturally occurring heat below-ground to create steam and generate electricity. They work best in locations near volcanic activity, such as Iceland and the mountain West of the United States. However, much of the global geothermal resource is deep dry hot rock that does not hold enough water to drive steam-powered turbines. That means water has to be added. Enhanced geothermal systems exploit the dry hot rock by injecting large volumes of water down into fractured rock from an external water supply at the surface. The injected water absorbs the geothermal heat and is pumped to the surface to power the steam cycle. The same water volume is then injected back into the rock to form a closed-loop system. However, because of lower operating temperatures and losses during the round trip from the surface, geothermal systems need more water than nuclear or solar thermal power plants.

Electricity generation from burning biomass like wood pellets or trash requires the use of similar amounts of cooling water as coal- and nuclear-fueled thermoelectric facilities, as the power generation process is very similar. Beyond the water needed to cool the power plant, water is also needed to develop the fuel.

The power sector is not the only part of the energy supply chain that needs water: water is also needed to produce fuels, including the growth, mining, extraction, and refining of fuels. There are several ways to think about the water intensity of fuels. We can think about the water needed per gallon of fuel. Or the water needed per unit of energy. But both notions are problematic. Failing to establish our metrics clearly can confuse public policymaking, which can lead to bad outcomes.

Speaking in rough orders of magnitude, it takes roughly one gallon of water per gallon of gasoline from conventional petroleum, one thousand gallons of water per gallon of ethanol from irrigated corn, and one hundred thousand gallons of water per gallon of biodiesel from algae.[14] While these ratios are easy to contemplate, they skip some key aspects. In particular, they do not consider the differences in energy density or the type of water. A gallon of ethanol has roughly two-thirds the energy of a gallon of gasoline, so treating both gallons the same is misleading. And algae can grow in saltwater, which is abundant, so maybe it does not matter how much water it requires.

And what about natural gas and hydrogen, two gases whose energy density per gallon varies for different pressures and temperatures? Pipeline-quality natural gas in the United States at standard conditions (standard temperature and pressure, which is zero degrees Celsius and normal pressure at sea level) has a density of about one million Btu per thousand cubic feet. Compressed natural gas is typically stored at a pressure 200 to 250 times higher than at sea level, which means compressed natural gas requires less than 1 percent of the volume it would require at standard conditions. Liquefied natural gas is cooled to −162 degrees Celsius, after which it occupies a volume that is less than one six-hundredth of gas at standard conditions. Comparing the volume of water per volume of natural gas can lead to ambiguities, as the volumetric energy density of natural gas varies so much, depending on its storage conditions. Hydrogen has a similar problem. Hydrogen has the same energy content as gasoline when compressed to seven hundred times the pressure at sea level. However, uncompressed, the same volume has a much lower energy content.

Finding comparable measurements complicates matters further: a "gallon" is a clumsy or irrelevant metric for electricity. Water matters for electric vehicles since they are becoming more popular and power generation requires so much water. But what is a gallon of electricity?

We could instead express water intensity in terms of gallons of water per unit of energy. Doing so is more precise than the volume-to-volume approach since the energy density variations from gasoline to ethanol, natural gas, and hydrogen can more easily be accommodated. The energy content of electricity can also be incorporated, despite having a nonobvious volume.

While this approach is indeed an improvement, it still falls short because it fails to include the variations in conversion efficiency for the different fuels in the engines and motors that drive our cars. An internal combustion engine has a typical efficiency range of 15–30 percent. Motors that propel electric cars have a typical efficiency of 80–90 percent. That means, even the water-per-unit-of-energy metric is not perfect, as it fails to capture the fact that electrical energy takes a car three times further than a gasoline car per unit of energy. It is better to think about the water needed per unit of energy service that is provided: per mile traveled in a car, or per unit of heating or lighting, for example.[15] Switching our thinking from miles per gallon (of gasoline) to gallons (of water) per mile would be a clearer indicator of the water intensity of transportation fuels.

Beyond these simple ratios, it is also important to distinguish the water needs at various stages of the life cycle of fuel production and use. These stages include production, upgrading, transport, and end-use. Production is the point of extraction for oil and gas; mining for coal and uranium; or photosynthesis for biofuels. Upgrading includes refining to turn crude oil into gasoline, jet fuel, or diesel; biorefining to turn corn or sugar into ethanol; and enrichment to turn ore or yellowcake into enriched uranium. For oil and gas, water is used to release the fuels out of the ground by waterflooding of conventional reservoirs, steam-assisted gravity drainage for the Canadian oil sands, or hydraulic fracturing of shale formations. Water is used again both as an input chemical and to make steam at the refinery. For biocrops, water is used for irrigation and to make steam for fermentation. For coal mining, dewatering often has to occur before the coal can be removed. Water is also used to control dust at coal mines. For uranium, water is used in mining to leach out the desired minerals and then to cool the power plants that provide the electricity for centrifugal enrichment.

In addition, water is used as a process input and a feedstock for process steam at refineries to upgrade the crudes into higher-value products. Typical volumes of water that are needed end-to-end for petroleum-based fuels from extraction through refining are approximately three to four gallons of water per gallon of fuel.[16] For natural gas, the volumes of water are approximately six to twelve gallons of water per gallon equivalent of oil.[17]

Generally speaking, unconventional oil and gas production sites are more water intensive than conventional oil and gas production. For oil sands production in Canada or heavy oil production in Venezuela, water is used to make steam to make the sticky, heavy oils flow out of the wells more easily. Water is also a critical input for hydraulic fracturing, during which jets of high-pressure water are injected into shale formations to cause fractures that increase the permeability of the reservoir.[18] Typical injection volumes are 2–9 million gallons per well of injected water.

Along with the millions of gallons of water, approximately one half million pounds of sand are included as "proppants" that hold the cracks open to increase gas flow.[19] In addition, chemical additives such as acids, surfactants, biocides, and scaling inhibitors are used to increase productivity. The typical composition of frac fluids is 98 percent sand and water, and 2 percent chemical additives. As producers become more water-efficient, using less water per well, the relative fraction of chemicals increases. That actually invites an environmental conundrum: using less water is an environmental objective, but the use of more chemicals makes many environmentalists uneasy. And higher fractions of chemicals, in particular gels and acids, are used for shale formations that produce a lot of liquids along with the gases.

One story about a boomtown in Texas recounted the following anecdote about what might be witnessed at a local bar: "Many's the time you'd see a man come in, order a quart of whisky poured in a bowl and go to washing his face and hands. Damned good reason for that: water had to be hauled miles and it cost like blue blazing hell. . . . It took many hundred barrels of water to drill a well those days . . . but water cost three dollars a barrel."[20] That story is not from the shale boom in the 2010s—it was about the oil boom in Burkburnett, Texas, written for a story in *Cosmopolitan* in 1939. Water has always been expensive for oil and gas booms, so the modern shale frenzy might not be that different after all.

A significant fraction of the injected fluids comes back out of the wells as wastewater, including drilling muds, flowback water (which is the portion of the injected frac fluids that are returned), and produced water (which is the naturally occurring water in the reservoir that gets brought up through the well). The volumes can range from 15 percent to 300 percent of the injected water, depending on the geological characteristics of the formation. That means some wells return more water than was injected, whereas other wells keep most of the water downhole. Overall in the United States, about seven or more barrels of water are handled, produced, or injected for every barrel of oil that is produced. That means oil and gas companies are really water companies who happen to have high-value byproducts, namely, the oil and gas.

Unfortunately, produced water usually has very high salinity and is difficult to treat. Underground injection into saline aquifers is one disposal method, though the water can also be treated and reused. In Texas, there are abundant injection sites, which makes disposal relatively easy. By contrast, in the Marcellus Shale region of the northeastern United States, the wastewater must be treated or trucked elsewhere as pipelines, treatment plants, and injection sites are limited in availability. The wastewater could also be shipped to more distant disposal sites, which requires energy for transport and introduces risks of accidental spills along the way.

In many cases, the produced water volumes far exceed the volumes of fuels that are produced, making wastewater disposal a potential constraint on production. Wastewater injection is common in areas like Texas and Oklahoma that have a lot oil and gas production. The waste fluids are injected at high pressures deep underground to keep them sequestered out of the hydrologic cycle.

However, doing so can induce seismicity, which is a scientific way to say "cause earthquakes." They happen when the wastewater, pumped at very high pressures and high flow rates, is injected at a fault, pressurizing and lubricating the fault, then triggering an earthquake. In 2014, because of significant oil and gas activity and wastewater disposal that goes along with it, the state of Oklahoma was the most seismically active of the fifty states, exceeding even California.[21] Unfortunately, those earthquakes take place in the middle of the country, where seismic codes for buildings are weak or nonexistent. It didn't take long for residents in Oklahoma and northern Texas to notice that the rise in earthquakes and nearby hydraulic fracturing were related, though regulators, under significant pressure from industry, seemed reluctant to connect the dots. In 2015, the U.S. Geological Survey linked those earthquakes to the underground injection of wastewater from oil and gas production.[22]

The earthquakes became so frequent, along with other concerns about the rise of shale production, that the citizens of Denton, Texas, a conservative city that is supportive of the oil and gas industry, overwhelmingly passed a referendum to ban fracking in the city in fall 2014. In response, in 2015 the Texas legislature, backed with a lot of campaign donations from the oil and gas industry, changed the laws to prohibit cities from passing such bans.

Buying, trucking, injecting, and disposing of the water costs a lot of money for oil and gas operators. A large drilling pad with multiple wells might cost $10 million to complete; up to 10 percent of the cost can be for water management. In an arid climate enduring an oil and gas boom, water is precious, so water prices can escalate. During the shale boom in the 2010s in Texas, water prices for hydraulic fracturing increased a hundredfold.[23]

The mining for uranium and coal can have significant impacts on water quality. Underground abandoned mines affect groundwater because water flows through the mines, picking up contaminants before traveling onward to aquifers. Surface mines, in particular mountaintop removal mines, also affect water quality because of the topographic disturbances that push soil and other mining residue into waterways. In addition, ponds and other impoundments are often used to store waste from coal-mining operation and ash from combustion at the power plants. When those impoundments fail, the pollutants move into waterways.

It is not just the fossil fuels that require water and impact water quality: biofuels have the same issues. Because biofuels are grown domestically and take carbon dioxide out of the atmosphere during growth, they have received a lot of policy support and interest. The most common forms of bioenergy include liquid fuels like ethanol for transportation and solid fuels like wood pellets for the power sector, though biogas (a renewable form of natural gas) can also be produced and used for those same applications. The water needs for growing biofuels vary widely depending on what is grown, where it is harvested, and whether or not it requires irrigation. Some biomass sources, such as forest trimmings and pulp and paper industry waste, use only natural precipitation for biomass growth. In contrast, dedicated energy crops and crop residues often come from irrigated lands with large volumes of human-applied water in addition to natural precipitation.

Although refineries need just a few gallons of water to produce a gallon of fuel from crude oil, biorefineries turning corn starch into ethanol consume three to ten gallons of water per gallon of ethanol, mostly as a source of steam for fermentation. Producing ethanol from Brazilian sugarcane consumes twelve to twenty-four gallons of water per gallon of ethanol.[24] While those numbers are greater than for petroleum-based fuels, this water consumption is just one part of the life cycle for biofuels. Notably, in the United States, fossil fuels are often used to create the steam at the biorefineries, which means biofuels are more carbon intensive and fossil-fuel dependent than many people might expect. Because biorefineries produce tens to hundreds of millions of gallons of fuel each year, they consume hundreds of millions of gallons of water per year, creating a localized impact and competition with other water users.

In addition to the water used for fermentation at the biorefinery, both irrigated and nonirrigated biofuel feedstocks need significant amounts of water to grow, on the order of several hundreds to low thousands of gallons of water per gallon of fuel. Whether the water is "counted" depends on whether the water was from irrigation. By convention, if the water was provided by rainfall, then governmental inventories do not include the water in their accounting. In actuality, the nonirrigated biofuels also need water for the growth phase, but we simply do not count it. However, that water is taken out of the system from other purposes and might also be worthy of tracking.

Unlike the oil and gas industry, which can use saline water for injection into wells, photosynthesis of traditional energy crops uses freshwater. For irrigated U.S. corn in 2003, the average irrigation withdrawal was nearly eight hundred gallons of water per gallon of ethanol. Soybeans are a little less water intensive than corn. For irrigated U.S. soybeans, the average irrigation withdrawal was approximately five hundred gallons of water per gallon of fuel. It is worth noting that trade associations and other advocates for the renewable fuels industry conveniently leave out the water intensity of crop growth and fossil-fuel dependence of biorefineries in their promotional materials, instead focusing on how biorefining is not that much more water intensive than traditional fuels.

Generally, the impact of growing biofuels is lower in water-rich regions, where irrigation is not necessary. For example, the vast majority of biofuels produced in Brazil are from rain-fed sugarcane, decreasing the irrigation water requirements for ethanol production. Brazilian ethanol production also uses the waste biomatter—bagasse from the sugarcane, rather than coal or natural gas as the fuel in the biorefinery—to ferment the sugar into alcohol. In addition, sugarcane cultivation in Brazil does not cause topsoil erosion the way corn does in the United States. In fact, sugarcane has been grown in the same location without loss of topsoil since the early 1500s, when it was introduced to Brazil by Martim Afonso de

Sousa. Sugarcane also produces more feedstock per acre, which means it needs less water per unit of energy produced. Overall, these factors suggest that sugarcane might be a sustainable option for Brazil.

Second- and third-generation biofuels, like lignocellulosic crops (switchgrass, wood chips) and harvesting of forest residues, are appealing because they presumably do not require irrigation, might be compatible for growth on degraded lands, and are not expected to trigger soil erosion. However, they might need much higher energy inputs at the biorefinery to convert from cellulose to starches than sugars. Algae has also been identified as a high-potential advanced biofuel. It needs even more water than corn, soy, or sugar, but that water can be saltwater or wastewater effluent, which means we might not care how water intensive it is.[25]

Getting energy to market also requires water, with barges along inland waterways and ships on the oceans moving crude and finished energy products from source to market. In particular, barges are used extensively for shipping coal and refined petroleum products, especially in the United States along the Mississippi River. China moves even more tonnage by its inland rivers. That means extended drought that lowers the water levels of the Mississippi or Yangtze Rivers puts supplies at risk for power plants that receive their coal by barge.

The oceans are also used for moving the world's fuels. Of the world's approximately 90 million barrels of daily oil consumption, 55 million are traded across country boundaries.[26] A significant fraction of that is shipped by supertankers over oceans, with the rest of it moving by pipeline. Each supertanker can hold up to 2 million barrels of oil. Of the 55 million barrels per day of petroleum trade across boundaries, about 38 million barrels of it is crude and the remainder is refined products. Moving oil by ship exposes the oceans to water quality risks from spills that can occur when ships run aground or are attacked. Because the petroleum is also liquid, it disperses quickly, spreading across large areas. Unfortunately, there have been many famous oil spill incidents over the years.

In addition, trillions of cubic feet of liquefied natural gas (LNG) moves each year across the oceans in specialized ships that have spherical containers designed to keep the liquefied gas cold so it stays in liquid form.

Before operators load the fuel onto the ship, multibillion-dollar liquefaction facilities that use a lot of energy cool down the natural gas to make it a liquid so that it can be exported more easily, as liquids are denser than gases. The ships arrive at multibillion-dollar gasification facilities designed for import. Along the way to its destination, some of the liquefied cargo boils off, and that gas powers the ships.

Once the LNG arrives at its destination, it might need even more water. Unlike power plants that use water as a coolant, large LNG facilities can use water as a source of heat. For example, there is a facility in the Adriatic Sea that takes very cold LNG from Qatar and turns it into gas. To do so, it uses the heat of the ocean. To make LNG, liquefaction facilities cool the gas to a temperature lower than −162 degrees Celsius, at which point the methane liquefies. Even though the ocean is cold, the LNG is even colder, and that means the water in the Adriatic Sea can be used as a source of heat for boiling the LNG from liquid to the gas phase. In this case the water gets colder as the methane gets warmer, and so instead of the risk that power plants have where the water can be too hot, the risk instead is that water is returned too cold. While regulators in the United States are concerned that power plants will return water to the environment at a temperature that is too hot, for these LNG facilities, regulators are concerned that water will be returned at a temperature that is too low to be safe for the aquatic environment.

Increasingly, coal is also moved by ships. Coal is historically produced and consumed within the same region. Compared with natural gas and petroleum, a smaller fraction of coal use is traded intercontinentally. However, coal export markets are growing. Because of the cheap gas from the shale revolution in the United States, the power sector's consumption of coal has dropped, displaced partly by natural gas. That has led to lower prices for coal, which make it attractive to European and Asian power plants, a new destination for coal mined in the United States. That means coal is mined, moved by train to ports, moved over the ocean on massive coal transport ships, before being unloaded at receiving terminals to be sent by train to the power plants. Surprisingly, coal can also be moved by pipeline.[27] While it is hard to imagine shipping solid materials by pipeline, by adding water, coal slurries can be produced that are fluid enough to transport by pipe. This approach requires mixing finely ground coal with significant volumes of water, up to hundreds of gallons of water per kilowatt-hour of electricity that is ultimately produced.

All of these transport modes use water, which means water is at risk from contamination or quality degradation from spills and accidents. And those transport mechanisms are vulnerable to drought and flood-related interruptions.

While the quantity of water needed for energy is extensive, water quality issues are also relevant. Energy can be used to improve water quality through water and wastewater treatment. But, energy can also degrade water quality, usually because of mistakes, accidents, or systemwide effects.

Oil spills are a particularly high-profile example of the risks to water quality. The oil spill discussed earlier from a blowout at a drilling platform in the Santa Barbara Channel was responsible for the release of 80,000 to 100,000 barrels (3–4 million gallons) of crude oil into the waters off the coast of California early in 1969. At the time, that was the largest spill in U.S. history. The Santa Barbara incident was surpassed in volume by the 1989 *Exxon Valdez* accident twenty years later, which spilled 11 million gallons of oil along Prince William Sound, Alaska. The explosion and subsequent oil spill from the Deepwater Horizon disaster at BP's Macondo well in the Gulf of Mexico on April 20, 2010, was even larger, unfortunately demonstrating that as oil production becomes more difficult, the scale of accidents can increase too. That accident is also a stark reminder that low-probability, high-impact risks of petroleum exploration in aquatic environments can lead to disaster.

All spills are bad news, but the spill from the failed blowout preventer at the Macondo well in 2010 was particularly gripping because it was aired on live television. It got so bad that commentators, including me, wondered out loud if detonating a nuclear weapon inside the well would help seal it off. That became a front-page headline on the *New York Times,* reminding me to never muse out loud or by email around a reporter.[28] Shortly after this idea got picked up by the news and became a punch line on the *Daily*

Show with Jon Stewart, I was vindicated by a distinguished veteran of the U.S. nuclear weapons testing program, who confided to me that he was upset I beat him to the punch as he was going to make the same recommendation publicly. He cited his experience from the weapons tests in the 1960s, where they would conduct underground and undersea explosions and so they knew exactly how tightly it would seal off a well. It turns out that President Nixon had supported the use of nuclear weapons for oil and gas production as a way to fracture wells decades before hydraulic fracturing in shales became popular, so we had more experience with that crazy idea than I anticipated.[29]

And while those three spills in the United States captured media attention here, many other spills happen worldwide. Other famous spills include the *Amoco Cadiz,* which broke in two off the coast of Brittany, France, in 1978, spilling 67 million gallons of oil.

While a large tanker carrying crude oil as a high-dollar freight makes big news when it loses its valuable cargo in such an environmentally risky way, in fact, over one hundred smaller spills happen each year that are hardly as newsworthy and are not from oil tankers. Though small, they can still have an impact on a local scale.[30] One of those events was in the San Francisco Bay when a cargo ship collided with the Bay Bridge in early November 2007, spilling 58,000 gallons of its bunker oil into the bay. And a pipeline that broke on land but along the coast of Santa Barbara sent thousands of barrels of oil into the water in 2015, reminding observers of the 1969 oil spill in the same region.

These water-related spills can also happen in urban settings. For example, in January 2008, an oil spill in my hometown of Austin flowed into Waller Creek, which is located downtown. As recounted by the local paper, it turns out that 8,000 gallons of fuel oil, used for on-site power generators, had been sitting idle in an underground storage tank for more than a hundred years in an alley next to the storied Driskill Hotel, where President Lyndon B. Johnson took Lady Bird Johnson on their first date (he asked her to marry him at the end of that day). Subsequent building owners on the neighboring lots forgot about the tank and paved over and built beside it for decades, not realizing it was there. Ultimately a large water leak caused water to spill into the tank, pushing out 4,200 gallons of fuel oil. So in this case it was water that caused the spill, and water that was impacted by the spill. Even this small spill had cleanup costs of approximately $200,000.

Hurricanes can also cause spills. Newspaper coverage of Hurricane Ike in 2008 reported that its "winds and massive waves destroyed oil platforms, tossed storage tanks and punctured pipelines. . . . At least a half million gallons of crude oil spilled into the Gulf of Mexico and the marshes, bayous and bays of Louisiana and Texas."[31] Industrial centers near Houston and Port Arthur and oil production facilities off Louisiana's coast were hardest hit by the storm. Over the span of a few days just before, during, and after the hurricane, 52 oil platforms out of roughly 3,800 in the Gulf of Mexico were destroyed, and at least 448 releases of oil, gasoline, and dozens of other substances were reported, impacting the ground, air, and water in Texas and Louisiana. Of those, "by far, the most common contaminant left in Ike's wake was crude oil." And that is just one storm. Hurricanes like that cause the kind of disruption to the oil and gas sector that terrorists could only dream of.

Spills can also occur from leaking pipelines. Unfortunately, spills have been part of the history of pipelines since the very beginning. The U.S. Department of Transportation keeps a record of what are termed "significant pipeline incidents"—those that cause fatality or injury, have costs in excess of $50,000, liquid releases of fifty barrels or more, or releases that cause fire or explosions.[32] On average, there are over 125 significant spills of hazardous liquids yearly in the United States, releasing an average of more than 120,000 barrels of hazardous liquids annually. Despite that record, pipes are generally considered a safer, cheaper, and cleaner way to transport liquids than trucks and trains.

However, with a growing industry to develop the carbon intensive and more corrosive Canadian oil sands, long-haul pipelines that bring that form of crude, known as dilbit (shorthand for "diluted bitumen"), across the United States to refineries are controversial. The Keystone XL pipeline, which can transport more than a million barrels per day, was opposed strongly in 2012–2015 by environmental groups and the Nebraska legislature, partly because of fears that a leak above the Ogallala Aquifer would cause irreparable harm.

Two prominent examples of serious leaks from pipelines carrying dilbit occurred in 2010 and 2011. One was the rupture of a thirty-inch pipeline operated by Enbridge in July 2010 near Marshall, Michigan.[33] That rupture released nearly a million gallons of crude into the Talmadge Creek, which then flowed into the Kalamazoo River. The spill was followed by heavy rains, which carried the oil even farther. *InsideClimate-News* won the 2013 Pulitzer Prize for national reporting for its coverage of the disaster. What is particularly vexing about that spill is that the heavy dilbit sank to the bottom of waterways, where it became submerged under the riverbeds and for which common cleanup techniques used for spills of lighter, conventional oil weren't as effective. The total costs for cleanup, which started in 2010 and required at least three years, were estimated to exceed one billion dollars, making it the costliest oil spill cleanup in U.S. history. By contrast, cleanup for the high-profile spill in the Yellowstone River in 2011, which released 63,000 gallons of crude oil, was easier, even though it contaminated seventy miles of river, because the oil was lighter and did not sink into the riverbed. That difference in the crude reduced the cost, lessened the impact, and simplified the cleanup.

Oil spills have the most visible and headline-grabbing effects on water quality, but natural gas also poses its own risks. It is difficult to differentiate natural and anthropogenic contamination, but there are concerns that ramped-up production will exacerbate the risks of the latter. While these risks are very real, there are aquifers that have naturally occurring natural gas. One notable example is the town of Burning Springs, West Virginia, which is named for the phenomenon that owing to the natural gas in the aquifer, the water can be lit on fire.

These water quality concerns are particularly prominent for unconventional gas production from shale formations using hydraulic fracturing because of the large volumes of water that are used for well completion and the large volumes of wastewater that are generated, and because drilling penetrates the water table. Notably, conventional drilling also punctures the water table and produces wastewater, so those risks aren't specific to unconventional production. At the same time, the volumes of water are different, the types of chemicals injected into and produced out of the wells are different, and the higher pressures used for hydraulic fracturing mean the risks for well failures might be higher.

These risks are very real. Yet even with all the attention to water quality that accompanied the shale boom, several questions remain. Which activity introduces greater risk: belowground well work (drilling and completion) or aboveground functions (trucking and storage in ponds)? Which belowground activity introduces greatest risk: drilling, hydraulic fracturing, or wastewater injection? Which water sources are at greater risk: surface water or groundwater? Which water type is the greater risk: frac fluids that are injected into the ground or the wastewater that comes out of the ground?

The concerns about the impact of shale production on water quality occur at several steps and locations in the shale production life cycle: at the water table, from migration underground, from seepages from the storage ponds, from the outputs of the wastewater treatment plants, from the wastewater injection, and from truck accidents that cause spills. Anytime the water table is penetrated by a well—something that has happened over a million times in the United States during more than a century of oil and gas production—there is risk of contamination. This risk is akin to the medical profession, where every time the skin is punctured to take a blood sample, there is a risk of infection. Done with the right precautions, the risk of infection is quite small, and the same is true with oil and gas operations: if companies case their wells in cement the right way—to the right depth, with the right quality of cement, letting it cure fully, and testing its integrity—then the risk of water contamination is quite small. But, mistakes have happened and do happen. They are rare, but impactful. And, if it's your water that is ruined, it's not very comforting to know that such a phenomenon seldom occurs.

There are also the concerns of chemical seepage from the point where the fracturing occurs to the groundwater. However, it is hard to imagine the chemicals flowing thousands of feet upward through the shale and other geological layers to the water table, when it would be so much easier to just flow through the well instead. Having said that, there are natural fissures and seeps, so it's possible the fracturing process activates those existing pathways through the rock.

While research to quantify the risks and impacts is still under way, anecdotally, the aboveground risks from spills, leaks from storage ponds, and truck accidents that cause a release of water from the tanks seem to be a much bigger overall factor.[34] And studies have clarified that the water quality risks are primarily from the drilling and cement work of the well, not the fracturing.[35]

The idea of water quality risks for our transportation fuels isn't a new one. For many years, MTBE (methyl tert-butyl ether) was a common additive to gasoline to improve automobile performance. However, it was prone to leaking out of underground storage tanks and trickling into the groundwater. After some litigation, the additive was eliminated from fuels in the first decade of the 2000s because of the risks to water quality. The loss of MTBE helped pave the way for a new, all-natural organic additive that promised to help boost the octane in cars without posing a risk to water quality: ethanol from corn. In fact, ethanol's potential as an additive is one of the reasons it received so much support in the form of tax credits, subsidies, and mandates in the Energy Policy Act of 2005 and the Energy Independence and Security Act of 2007. But corn ethanol poses its own risk to water quality.

Quantifying the water quality impacts of the agricultural portion of the biofuel life cycle presents new challenges because most of the impacts spread out over a large area, like the pollution from car exhaust all over a city. Pollutants are transferred into water by means of excess irrigation, rainfall, or snowmelt that flows over and through the ground as runoff, collecting manmade pollutants from farms—the chemicals used to fertilize the crops—as it moves. Since pollutants transferred to water bodies from contaminated runoff or percolation through the ground cannot be attributed to discrete sources, this type of water pollution is much more difficult to measure and regulate. Even though the connection between agricultural activity and nutrient runoff to downstream water bodies is widely accepted, pollution from agricultural sources is largely unregulated. The same story is generally true when it comes to the air quality impacts of farms, too.

Because of biofuels mandates, domestic production of biofuels—and the water pollution caused by farm runoff—has gone up.[36] Although all fertilized crop production loses nutrients, corn is particularly inefficient: it uses only 40–60 percent of the nutrients delivered to its roots. That means the rest of the nutrients run off into the neighboring ecosystem. In particular, there have been increases in nitrogen and phosphorus agricultural chemical concentrations and hypoxia (a "dead zone") in surface waters draining from farmland in the Mississippi River basin, and groundwater near farmland, into the Gulf of Mexico. This increase in nutrient loading from crop production has contributed to the growth of a large hypoxic area in the Gulf of Mexico, which is currently the second largest hypoxic zone in the world after the Baltic Sea.

It is hoped that other forms of bioenergy such as switchgrass and woody materials would not only lower irrigation needs but also require fewer agricultural chemical inputs and therefore have lower water quality impacts. Those prospective benefits are part of the motivations for the use of plants other than corn as a source of fuels.

As countries shift from conventional fossil fuel production toward unconventional fossil fuels and biofuels, the nature, extent, and location of water use and water pollution will be different. Consequently, the existing regulatory frameworks for protecting water quality may need to be updated and revised.

FIVE

Energy for Water

JUST AS WE NEED SIGNIFICANT volumes of water for energy, we also consume a lot of energy for water. We use energy to store, pump, treat, move, clean, and use water. Much of the water we use is for irrigation or power plant cooling. Usually the water for irrigation doesn't need to be treated beyond simple filtering, if at all. Power plants will treat water to prevent minerals from accumulating inside equipment, but they do not need the water to meet potable standards. Drinking water for our municipal systems typically requires extensive treatment. Then, after we use the water in our homes or businesses and flush it down the drain, it has to be treated again to raise it to a standard that is safe to return to the environment without causing damage to the ecosystem. The full supply chain includes pumping and conveyance from the original water source, water treatment, water distribution, end-use, wastewater collection, wastewater treatment, then discharge. Generally speaking, the treatment steps are very energy intensive, although there is great variability. Despite their requirements, heating at the end use is the biggest energy consumer of all.[1]

Though abundant globally, water is often out of reach. When water is far away from where we live or is dirty, we spend a lot of energy moving it, cleaning, or storing it. The idea that energy is needed for pumping water is certainly not new. The ancient Egyptians widely used Archimedes' screw, a clever device that uses the manual turning of a screw to elevate water. The tight coil of connected blades could keep raising water as long as it was continually operated. Its invention is attributed to Archimedes in the third century BCE, but it might have been in use even earlier. It is still employed today, in modern water-lifting stations at amusement parks, water treatment plants, and elsewhere. It turns out that robust designs are still useful thousands of years later.

The reason that water needs so much energy for pumping is because it is so dense: it weighs 8.34 pounds per gallon. That density is also one of the reasons why water is so valuable as a coolant and a process material. That also means it takes a lot of energy—and therefore money—to bring that water uphill. Marc Reisner's classic book *Cadillac Desert* captures this idea from the people living in deserts who begrudgingly noted that water moves uphill toward money and power.[2]

The energy needed for pumping water depends on how far the water needs to be raised, the rate at which it is raised, pipe diameter, friction, and so forth. The energy that is needed to raise water up out of a well is the energy that is required to overcome the force of gravity, which wants to pull the water back down. Raising a 2.75 gallon bucket of water (about 10 liters) a distance of about 330 feet (100 meters) requires approximately ten Btu (or 10,000 Newton-meters) of energy. Pumping the water up only 33 feet (10 meters), for example by raising the water from a river to the top of a nearby riverbank, requires one Btu (or 1,000 Newton-meters) of energy. That doesn't seem like much, but for large volumes, the energy adds up, and the distance it has to be elevated drives the energy needs.

For continuous flows, it is more useful to look at the power that is needed. Pumping 2.5 million gallons per day (29 gallons per second, or 110 liters per second)—enough water for 1,900 average Americans—out of an aquifer 330 feet below the ground requires 107 kilowatts of pumping power. Keep in mind that a typical house needs 1–3 kilowatts of power on average to run the whole place, so a pump that size consumes the same power as approximately thirty to one hundred homes.

A medium-sized U.S. city with a million residents might need about 150 million gallons of water per day. Raising that water from surface sources to elevated water treatment plants over a height of one hundred meters requires a little more than 6 megawatts of pumping power. Massive wind turbines are approximately 1 megawatt apiece, so that city would need a half-dozen of them running full-bore just for pumping water up from its source to the top of the hill, after which it can flow downhill to customers.

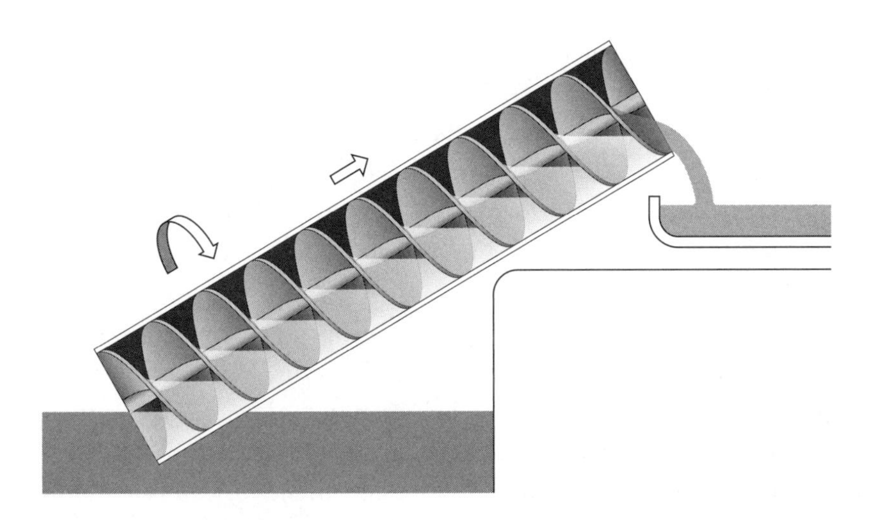

It takes a lot of energy to lift water. The Archimedes screw (*top*) is one mechanism for doing so. Energy is invested to rotate the screw, which raises the water, as at Schlitterbahn, a water park in New Braunfels, Texas. [Photo by Jeffrey M. Phillips]

That means pumping water uphill requires a lot of electrical power, which requires a lot of money. And money is a proxy for political power. So Reisner's book was correct: water moves uphill toward power and money.

Imagine pumping that water by hand, the way it once was. This energy intensity is one of the reasons why people like shallow wells and surface water.

And all those energy and power requirements for pumping water to higher elevations don't include the energy for treating the water to make it potable or pumping the water through pipes to our homes. For cities that have water treatment plants at high elevations, much of the distribution can be done by gravity, but not all cities have that advantage. Cities whose layout does not accommodate gravity-fed distribution with natural geography instead pump the water into ubiquitous elevated tanks that look like giant golf balls sitting on a tee.

In addition to the energy to convey the original source of water to the treatment plants, the treatment processes themselves require a lot of energy. Water and wastewater treatment require that energy for a variety of actions, including pumping, blowing, aeration, ultraviolet lamps, stirring, and the embedded energy in the chemicals that are added to the process. Words like flocculation, filtration, carbonation, and sedimentation are standard parts of the water treatment engineer's lexicon.

The amount of energy needed to treat water and wastewater to a suitable form depends on a variety of factors, such as how contaminated the source water is, the nature of the contamination, what the water will be used for, plus the physical features and treatment approach of the facility. Dirtier water generally requires more energy for treatment, and end-uses that require high standards of cleanliness also need more energy. Hospitals, semiconductor cleanrooms, and food preparation facilities need water that is much cleaner than what is required for cooling industrial equipment or irrigating farms. That means there is great variability nationally for the energy intensity of water treatment.

Bottled water is perhaps the most energy intensive use of all.[3] Energy is required to process, bottle, seal, and refrigerate the water, but it turns out that the energy that goes into making the plastic bottle itself is the biggest piece. And, if that bottle is moved by trucks, planes, or ships over long distances, the energy consumption goes even higher. In modern cities that have well-functioning piped water systems that draw from local sources, the water is produced in a comparatively energy-efficient and affordable way. In those situations, the extra money and energy for bottled water seems like a wasteful indulgence that offers little additional value. However, after a natural disaster such as a hurricane wipes out a local water system, or in developing countries where water systems are contaminated or do not exist at all, bottled water can be a lifesaver.

Considering all the energy used for water treatment, the national average is 1,400 kilowatt-hours per million gallons for sourcing, treating, and distributing surface water.[4] The water used for electricity is often denoted in gallons per kilowatt-hours, while the energy for water can be tracked in kilowatt-hours per million gallons. Energy for conveyance varies from zero for gravity-fed systems, as in New York City, to approximately 14,000 kilowatt-hours per million gallons for Sierra snowmelt water delivered via the State Water Project from Northern California over mountain ranges to Southern California destinations such as San Diego.

It turns out that the State Water Project, California's ambitious water conveyance project that moves water by aqueducts across the region, singlehandedly consumes 2–3 percent of the state's electricity.[5] The project would make the Romans proud. The world's largest water pumps were invented just to lift the water two thousand feet over the Tehachapi Mountains. That shipment of water is still a source of contention within California today. The total length of the State Water Project is about seven hundred miles, serving 25 million people and consuming 5 billion kilowatt-hours each year. The Central Valley Project has about five hundred miles of canals, consumes 1 billion kilowatt-hours annually, and includes eleven hydroelectric power plants to harness electricity from the snowmelt. A lot of that water goes to Southern California cities plus agricultural operations growing nuts and fruits. During the peak of the California drought in 2015, it was a widely noted complaint that it takes a gallon of water to grow an almond and several gallons to grow a walnut in the Central Valley of California, which provides nearly the entire nation's supply of domestic nuts.

When the water is sourced from groundwater, the national average is higher, at 1,800 kilowatt-hours per million gallons for sourcing, treating, and distributing it, primarily because of the additional energy for pumping the water to the surface. Brackish groundwater requires even more energy, with a range of 3,900–9,750 kilowatt-hours per million gallons. That range depends on the level of total dissolved solids in the water: the more salts that have to be removed, the more energy is required. At the high end of the scale is seawater desalination, with a range of 9,780–16,500 kilowatt-hours per million gallons. Desalination is very energy intensive, and its varying energy requirements are a function of water temperature and salinity. Although desalination systems are robust and drought-resistant because the world is awash in saline water, they are very energy intensive. In addition, they produce brine streams of water that are even saltier and require disposal. Both the energy consumption and the disposal represent environmental impacts that are important to consider.

There are basically two standard approaches to desalination that are widely implemented: thermal systems and membrane systems. Those two approaches provide about 95 percent of all desalination globally, with other minor approaches such as freezing and electrodeionization providing the rest.[6] Thermal systems include actions as simple as boiling water, which separates out the salts by evaporating the water, which can subsequently be condensed. This is just distillation as learned in high school chemistry. This system is very effective but requires a lot of energy. Thermal approaches also include more sophisticated systems like multistage flash, which uses a series of steps and pressure drops to evaporate water, and multieffect distillation. Thermal systems can be very crude, as they can use any source of heat. This crudeness means they have low efficiency in practice, but it also means they can use waste heat as the source. So when thermal desalination systems are integrated into large facilities with a lot of waste heat, such as power plants, then the overall efficiency is improved. Membrane systems use a semipermeable barrier to filter out the dissolved solids. The most popular membrane approach is reverse osmosis. Other

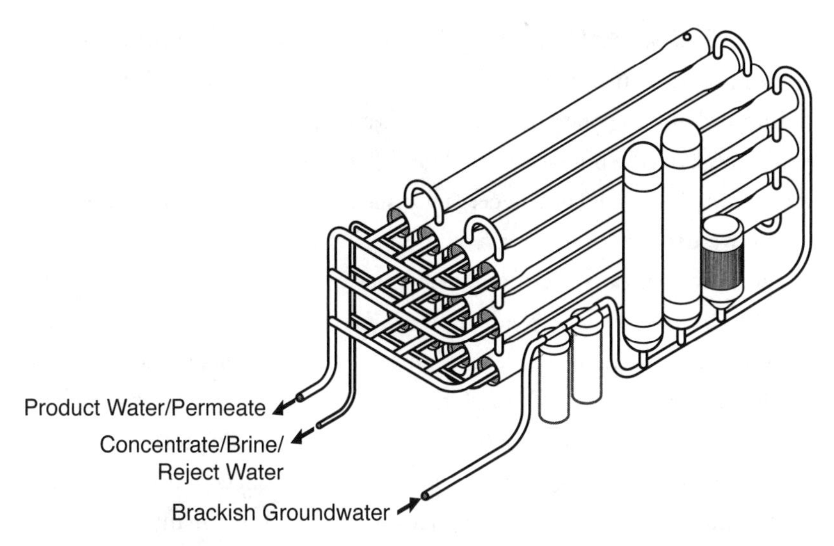

Many desalination facilities use reverse osmosis, pushing salty water through cylindrical membranes that separate the salt from water. Doing so creates two outputs: a stream of less salty water (permeate, or product water) and a stream of saltier water (concentrate, brine, or reject water).

membrane approaches include forward osmosis, nanofiltration, and electrodialysis.

Reverse osmosis filters use electrically driven pumps to push water through membranes that separate out the salts. The membranes separate two streams of water with different levels of salinity. Normally, osmotic pressures would drive salts from the side with greater salinity toward the side with lower salinity as a way to achieve equilibrium, so that concentrations on both sides are about the same. The pumps reverse this direction, causing the salty side to get saltier and the fresh side to get less salty. Reverse osmosis systems are generally more energy efficient than boiling water, but are expensive technologies that require electricity as opposed to just heat.

After treating water and getting it to our homes, we use it to flush away our wastes. In a book aptly titled *The Big Necessity: The Unmentionable World of Human Waste and Why It Matters,* the author Rose George humorously describes her adventures going around the world and writing about how different people relieve themselves.[7] She examined the practice in the United States, United Kingdom, Japan, China, and India, revealing some interesting insights along the way. For something so basic to human nature as emptying one's bowels, there is a surprisingly varied range of approaches. Even the phrase as simple as "going to the bathroom" is uniquely American, as we do not necessarily have baths in most of the rooms where we use the toilet. In the United Kingdom and Europe, the bathroom actually means the room with a bath in it, so instead they use the word "toilet" or "water closet" or "loo." For the United States, the "toilet" is the actual commode, whereas in Europe, "toilet" refers to the room that holds the commode. The French have the phrase "toilet water" to denote perfume. The Japanese have the world's most luxurious toilets, with robotic functions, heating, built-in cleaners, and all sorts of options. Many rural Chinese have outhouses or latrines. And in India, many people defecate openly in the fields.

In addition to observing the differences in the act of defecation, Rose George also follows where the waste goes. She travels down into the sewers and to the wastewater treatment plants to see the technology and the investments required by society to make them work. The understated conclusion from all of her writing is that waste matters. Waste is a differentiator of society. Societies that deal with their waste are healthy and wealthy. Societies that do not manage their waste are sick and poor. The cause and effect is not clear: whether societies are healthy and wealthy because they manage their waste, or whether they manage their waste because they are healthy and wealthy is not obvious, but the correlation is hard to miss.

Managing wastewater and sewage is a big task, and it is all part of what we term "sanitation." The system is pulled together by sewers or drainage pipes that bring the solid-laden flows to the wastewater treatment plant. Sewers themselves can be a massive infrastructure undertaking, but then the wastewater treatment plant itself is where the real hard work begins. As my colleague Ashlynn Stillwell likes to say, "When we flush waste 'away,' it's not gone. Wastewater treatment plants are the 'away.' It's that place no one likes to think about. It's where all our waste ends up." When we took our students on a tour of a wastewater treatment plant, we told them we were taking them away. In the animated movie *Flushed Away,* the setting is the historic sewers of London, further reinforcing this notion that away and sewers or wastewater are synonymous. Given our distaste for our own waste, "away" is an apt description for where we want it.

Wastewater treatment plants are usually at a low elevation so that the sewage can flow by gravity to the facility. Consequently, most of the energy requirements are used for the treatment process as opposed to conveyance. On average in the United States, wastewater treatment requires 955–1,900 kilowatt-hours per million gallons, depending on how many stages of treatment are applied and the level of cleanliness.[8] However, based on local prevailing standards of treatment, the requirements can be higher. In Austin, Texas, the wastewater treatment plants require 2,700 kilowatt-hours per million gallons. One reason is that Austin is the only city in Texas, and possibly one of only a handful in the United States, where the river segment downstream of a major metropolitan area has a better water quality designation than the segments upstream. Raj Bhattarai, one of the principal engineers at Austin Water Utility, beams with pride when he gets a chance to tout this surprising fact. As a native of Nepal high in the Himalayas, he understands the importance of protecting watersheds for the benefit of people downstream.

What this all means is that the water system overall requires a lot of energy. And for municipalities, it is usually the single largest energy tab for the city government because water and wastewater treatment facilities are usually municipally owned. It is typical for a water and wastewater utility to consume 50 percent of all the electricity used by a city's government. The energy intensity of water and wastewater treatment remains an opportunity for conservation.

One key conclusion from all of these energy requirements is that treating water, treating the wastewater that is produced, then treating the effluent from the wastewater treatment plant again to potable standards is less energy intensive than desalination. Closing the loop and using treated effluent is one option for reducing the energy requirements for the water system. This process is called "toilet to tap," and it grosses out a lot of people. While the idea of toilet to tap may be distasteful, it is a fine way to provide water.

That approach is already in action indirectly when we discharge our effluent to a river or aquifer for the next city downstream to drink. Andy Sansom, a noted water expert, is fond of saying, "We are all downstream from somebody." Houston is drinking the wastewater from Dallas. Belgrade is doing the same with wastewater from Vienna. But somehow we feel better when the water goes through "nature" first, as opposed to staying in pipes the whole time. While nature does provide some cleaning services in many instances, it is not obvious that nature's cleaning is better than what our engineered facilities can achieve.

"Toilet to tap" works. Singapore built such a facility in 2000, calling it NewWater. The system provides 60 million gallons per day or 30 percent of the drinking water from reclaimed wastewater.[9] NewWater has worked fine and is slated to triple its capacity by 2060. Water recycling and toilet-to-tap treatment systems are also valuable for the military. Shipping water to the front lines is an expensive and deadly proposition. The long supply convoys span thousands of miles and are soft targets for enemies, meaning that the cost of water per gallon ultimately is several orders of magnitude higher as delivered to the military theater than at the local grocery store or from our taps. Consequently, the U.S. military spends a great deal of money testing and deploying on-site water treatment systems that can make potable water from degraded streams. These systems cost a lot of money and require extra space up front to be shipped to the forward operating base, but after that, they spare the need for a lot of shipments of water, saving lives.

The International Space Station also has a reclaimed water system to produce drinking water because the cost of shipping water to space is exorbitantly expensive.[10] It costs from $10,000 to $90,000 per pound to ship cargo into space. That means shipping freshwater to astronauts in space costs anywhere from just under $100,000 to nearly $750,000 per gallon. Because of the high costs for freshwater, the space station collects and treats the graywater from washing, urine, and condensed moisture from breath and sweat to be drunk again. The station does not have any blackwater, as the astronauts do not use toilets.

Interestingly enough, my research at Stanford University for my Ph.D. was a part of this project to create an onboard water treatment system for the space station. As a graduate researcher in mechanical engineering I invented and deployed laser-based sensors that could measure a variety of trace gases. One of my patents from that work was for a sensor that would measure very small concentrations of ammonia in the presence of other species, such as water vapor or carbon dioxide.[11] I also studied combustion, the cornerstone of our modern energy system, giving me an early insight into energy conversions, consumption, and impact from the perspective of a thermoscientist. I used the sensors I invented to measure emissions of pollutants in the flue gases from combustion systems. In particular, I looked at the unwanted emissions of ammonia that slip out of the smokestack after the ammonia had been injected to scrub pollutants out of the stack.

Smokestacks have a lot of carbon dioxide and water vapor. The ability to measure trace quantities of ammonia in the presence of carbon dioxide and water was a useful performance advantage. Just like smokestacks, humans breathe out a mixture of carbon dioxide and water. That means the ability to measure ammonia in a smokestack is good preparation for measuring ammonia on the space station in the air mixed from the exhalations of astronauts.

At the time, NASA was developing an on-board water treatment system that would turn the graywater into drinking water. I flew to NASA's Johnson Space Center with my ammonia sensor in hand to make some measurements of their space-bound water treatment system that was undergoing ground testing. Bringing the whole energy-water nexus story full circle, Johnson Space Center was named after Lyndon Johnson in honor of his enthusiastic support for the space agency while he was vice president in the early 1960s. This is the same President Johnson who pushed for the Rural Electrification Act that helped poor families far from cities pump water into their homes, much as the "Farm Woman's Dream" poster envisioned.

It was through this experiment that I teamed up with professors, postdocs, and students at Rice University. Dr. Frank Tittel was the principal investigator of the project and the person who extended the invitation to me for the collaboration. Through that project I interacted briefly with Dr. Bob Curl, who won the Nobel Prize with Rick Smalley at Rice University. This is the same Rick Smalley whose top ten list of grand challenges facing humanity included energy and water at the top. Unfortunately, I never met Smalley.

NASA had set up a test station with a large bioreactor that would convert streams of wastewater laden with organic and nitrogenous components into fresh, potable water. Making freshwater available onboard for the astronauts is a key life support mission and had remained a vexing challenge for decades, so this experiment was considered an important priority. We were tucked into a spacious, nondescript experimental facility on the campus with large pieces of equipment next to a bathroom. NASA employees, including scientists and support staff, were invited to use the urinal, sink, and shower to provide the sample wastewater materials the bioreactor would convert into drinking water. It was not unusual for commuters to ride their bikes to work in the morning and then hop into the experimental shower to freshen up before they started working. Then there was a steady stream of volunteers throughout the day who came by to use the urinal or toilet.

The blackwater from the toilet was sent to the conventional wastewater system, but the graywater was sent to the experimental water treatment system that we were evaluating. Because there was no blackwater to manage, the treatment requirements were less intensive. However, the graywater still had a lot of organic matter from the soap, skin cells, and dirt, and nitrogeneous matter from the urea in the urine. Furthermore, since sickness in space is hard to manage, the water standards still had to be stringent to avoid propagating any waterborne illnesses.

The water treatment system was a simple configuration of a glass tube vertically aligned with ceramic shells inside, which the graywater would flow through. The shells would accumulate some of the undesired contaminants and would force mixing with the treatment chemicals. The treatment process would produce off-gassing of water vapor, carbon dioxide, and ammonia. The carbon dioxide was from the organic contaminants and the ammonia was produced from the urea and other nitrogenous

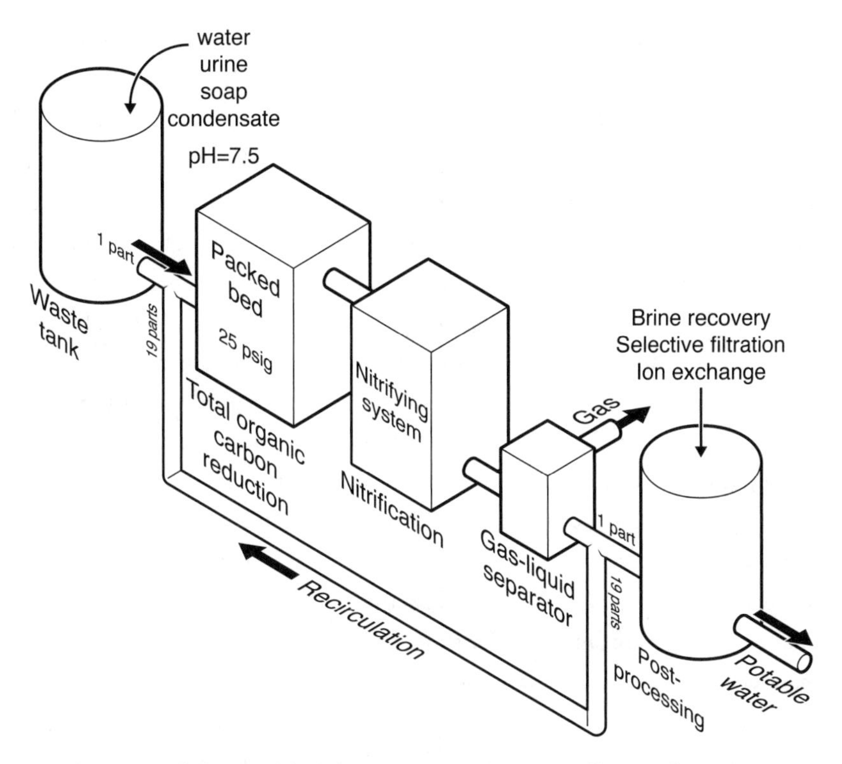

A schematic of the NASA Advance Water Recovery System for water processing aboard manned spacecraft. I invented an ammonia sensor for my doctoral dissertation that was used to measure off-gassing from the biological waste processor.

compounds. By measuring the fraction of ammonia mixed in with the water vapor and carbon dioxide, I could help the scientists assess the relative health of the bioreactor.[12]

I spent two weeks at NASA in August 2000 integrating my sensor into the experiments and taking data. With that experiment complete, I defended my dissertation less than three months later. One decade after I conducted those experiments, the water treatment system was finally launched in 2010.

While the energy needs for pumping and treatment are significant, it turns out the energy needs for preparing water at the end of the pipe are even larger. These energy investments are for processes such as filtering, chilling, pressurization, deionization, and heating. In particular, water heating is very energy intensive. It shouldn't be a surprise. Water requires so much energy for heating that it is used as the standard for energy units. In English units, one Btu (British thermal unit) is the energy required to raise the temperature for one pound of water by one degree Fahrenheit. In standard international units, one joule (J) is the energy required to raise the temperature for one gram of water by 0.24 degrees Celsius.

And water heating is a useful application for energy. After all, hot water serves many helpful purposes, including disinfection and sterilization at hospitals and washing dishes at home. Not to mention that hot showers (and therefore hot water) are a lot more comfortable than cold showers. So water heating is an important proxy for quality of life.

Including all of these end uses drives up the total energy consumed for water dramatically.[13] By adding up all these different uses of energy for water, appliance by appliance and sector by sector, a total national estimate for energy consumption for water can be deduced. Dr. Kelly Twomey Sanders, a professor in civil engineering at the University of Southern California, did just that. In fact, she was the first person to tackle that research problem on the national scale with so much accuracy. She found that the United States consumed approximately 12.3 quadrillion Btu of energy just for water in 2010. That year, the United States consumed 98.0 quads of primary energy, which makes water directly responsible for about 13 percent of consumption. Another 34 quads was consumed to generate steam for indirect purposes, such as process heating, space heating, and electricity generation. That last one is particularly impressive: we spend a remarkable amount of energy each year just boiling water at power plants to make steam that spins turbines to make electricity.

Of the 12.3 quadrillion Btu we consumed as a nation for water services and direct steam use, a little more than a fifth was just for heating water in our homes and businesses. On-site water pumping was relatively low in the residential sector in comparison to the industrial and commercial sectors, as housing units tend to be smaller. Residential water systems for most single-family residences operate off the prevailing pressure of the water distribution network, so pumps are seldom needed at all. But, for tall residential buildings, water must be elevated to tanks at the top of the building after which it is fed by gravity to the individual units. Those residential buildings need on-site pumping to raise the water. Large industrial facilities also require large quantities of energy to move water around on-site.

The total amount of energy we spend on water is impressive. It means as a nation the United States spends more energy on water than for lighting. And therefore the energy embedded in water is big enough to care about. The Natural Resources Defense Council and Pacific Institute referred to this phenomenon in their landmark 2004 report for California as "Energy Down the Drain." Literally, the energy embedded in cleaning, pumping, and heating water gets lost down our drains.[14] It also means that water conservation is a reasonable pathway to energy conservation. But interestingly, water conservation has been mostly ignored, perhaps because water is so cheap. By contrast, we have spent a lot of political energy as a nation fighting about lightbulb standards. President George W. Bush signed into law the Energy Independence and Security Act of 2007, which mandated an efficiency standard for lightbulbs. The standard basically prohibits the continued sale of conventional incandescent lightbulbs, whereas compact fluorescent lightbulbs and light-emitting diodes meet the requirements. Years later the standards remained a contentious issue and were removed by congressional Republicans as part of budget negotiations to keep the government open in December 2014. Similar fights have been waged over fuel economy standards for automobiles. The whole time, water has been ignored as an option for energy savings, despite its large energy footprint and significant opportunity to enable energy conservation.

SIX

Constraints

When the well's dry, we know the worth of water.

—*Poor Richard's Almanack*, 1746

ONE OF THE MAIN CONSEQUENCES of the interdependence of the two systems is that water constraints can become energy constraints and energy constraints become water constraints. Because the power sector is so water intensive, it is particularly vulnerable to water constraints. In fact, water can be too hot, too cold, too abundant, or too scarce for full operation of power plants, leaving a sweet spot where the water needs to be just right. That means heat waves, freezes, droughts, and floods all cause problems for power plants.

Heat waves (disregarding the drought that often accompanies heat waves) hamper the power sector in two primary ways: they reduce performance because of reduced power plant efficiency, and they put power plants at risk of violating thermal pollution limits. As an unavoidable consequence of the Second Law of Thermodynamics, power plants and automobile engines are less efficient when it is hot outside. That means the hotter temperatures of the heat wave degrade the efficiency of power generation. The efficiency of a power plant drops a percentage point just for the case where the water is ten degrees Celsius hotter from a heat wave. That might not sound like much, but for a nuclear power plant with one gigawatt capacity, a 1 percent loss of efficiency can cost more than five thousand dollars per day in lost revenues, because there is less electricity to sell.

In addition to the performance problems, environmental regulations have an effect as well. Because power plant cooling structures in the United States are regulated by thermal pollution standards from the Clean Water Act, there is a limit on the maximum exit temperature allowable for the return water from a power plant's cooling system. Typically those require-

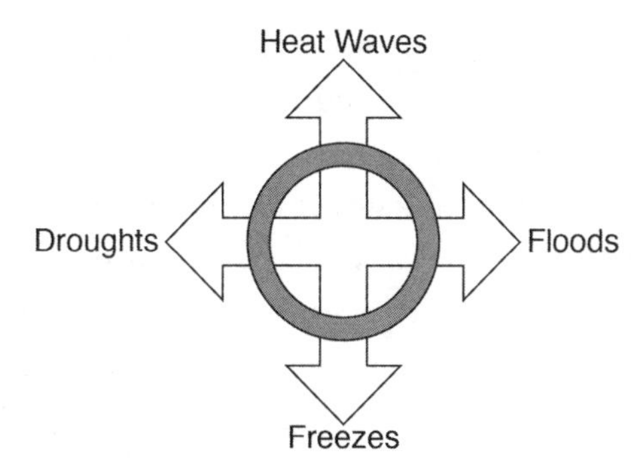

Extremes in water temperature and availability do not work well with power plants. The power sector needs water to be in a sweet spot—not too hot, too cold, too scarce, or too abundant—to function at full power.

ments, which are designed to protect the aquatic environment, are fixed at a particular temperature, say 104 or 112 degrees Fahrenheit. In some cases, the maximum allowable return temperature for the cooling water is based on the difference between the inlet and outlet. When a heat wave occurs, the temperature of the incoming cooling water will increase above normal. This puts the power plant at risk of returning water to its source at a level that exceeds its thermal pollution threshold, so it must intentionally cut back its operations to avoid a violation.

Extreme heat waves that affect the power sector can happen. The 2003 heat wave in France was one prominent example. That heat wave stretched across Europe, but France bore the brunt of its impact. In many places, the temperature anomalies in France were ten degrees Celsius (eighteen degrees Fahrenheit) hotter in 2003 than in 2001.[1] That was a killer heat wave, showing some of the health risks from climate change. According to one scientific paper, the heat wave killed approximately 70,000 people in Europe.[2] France in particular suffered nearly 15,000 to 20,000 additional deaths. In particular, the elderly were vulnerable. The demand for power was spiking as people turned on their air conditioners to avoid dying from the heat.

At the same time that the demand for power was peaking, the French nuclear fleet had to draw down its output to avoid violating the thermal pollution limits.[3] Because the French power sector has a very high contribution from nuclear power with water-intensive open-loop cooling systems and because the French nuclear plants are often sited on inland rivers (thirty-seven of their fifty-eight nuclear power plants are so situated), they are doubly exposed to heat waves. By contrast, ocean water maintains a much more stable and cooler temperature compared with inland rivers and lakes.

Because of the combination of the higher energy demand and higher river temperatures, as many as seventeen of France's fifty-eight nuclear reactors were at risk of violating the thermal pollution limits and had to reduce their output or turn off completely.[4] The rivers were too hot and the water levels too low to guarantee adequate cooling of nuclear power plants, putting the entire system at risk of failure. Just as demand for electricity was peaking for air-conditioning in response to the heat and with life hanging in the balance, power supplies were being cut back because of the very same heat wave. Ultimately, Electricité de France (EdF), the main power provider in France, requested exemption from its operational limits and cut its power exports in half to keep the power system operating. After those temporary exemptions were granted, nuclear facilities were allowed to operate but at reduced capacity. In total, the nuclear power fleet had to dial back its output by up to 15 percent for five weeks. France and many other regulatory bodies in Europe "overrode their environmental laws and allowed for higher waste water temperatures." When making the choice between impinging the environment through overheating of the ecosystem and saving human lives, decision-makers selected the latter.

Though the heat wave in France was certainly not a normal situation, the fear with climate change is that those killer heat waves will become the new normal. And, the phenomenon is not restricted to France: because of unseasonably warm weather conditions in May 2012, the Illinois Environmental Protection Agency (IEPA) granted Exelon's Quad Cities nuclear power plant a provisional variance from its temperature limits.[5] As in

France, the thermal pollution limits were temporarily voided during a heat wave to keep the power on.

Ocean water is generally cooler and its temperature more steady than inland rivers, but even seaside power plants are at risk of shutting down from heat waves. A few months after the Quad Cities plant needed a variance, the Millstone Power Station, a nuclear power plant in Connecticut on Long Island Sound, shut down for two weeks because of overly hot seawater, which exceed the 75-degree Fahrenheit temperature limits at the inlet of the cooling water system.[6] As before, the authorities did their best to shunt aside environmental regulations to keep the power on, but despite an emergency license amendment from the U.S. Nuclear Regulatory Commission to show a lower temperature by averaging different measurements, the safety threshold was violated, triggering a full shutdown.

Even though water can be too hot—a problem that might be more prevalent with global climate change—water can also be too cold. It is known by power plant operators that weatherization and preparing for cold storms includes steps such as draining nonessential water systems to avoid damage from ice formation, and confirming that the water in essential water systems was actually circulating, which reduces the likelihood of a freeze. So in effect, power plant operators know that freezing water poses a risk to reliability.

In an illustrative incident, Texas endured a statewide freeze in February 2011. That cold snap caused water to freeze in some instrumentation pipes, tripping some large coal plants off-line, which triggered a cascading series of power plant failures elsewhere, ultimately leading to rolling blackouts across the state.[7] On one day, six more coal-fired units at four locations went off-line. Then another old coal-fired plant went down when "an old switch filled with water and froze, causing it to break," adding to the strain.

Because the natural gas demand was at a record high to meet the heating requirements for homes and business, backup power plants fueled by natural gas did not have enough gas available in the pipelines. Ultimately, more than fifty gas plants were not able to work. Compounding things, many gas-producing wells use electrical equipment, and some gas compressors along pipelines use electrical pumps. When the power went out, those wells and compressors stopped working, making it even harder to meet demand and keep the pipelines filled with gas, causing pressures to drop and straining the system further.[8] Even if the pumps had not turned off because of the power outage, some gas wells turned off anyway because of a phenomenon known as "freeze-offs," which occurs when the water produced alongside natural gas freezes or crystallizes, blocking off the gas flow, effectively shutting-in the well.

Ultimately, between February 1 and 4, 2011, more than two hundred individual generating units within the Texas grid experienced an outage, were derated, or failed to start.[9] All of these events combined triggered statewide rolling blackouts: since enough power could not be produced to meet demand, demand had to be cut across the state. The grid operator, sitting in a control room, starting turning off the power to different parts of the state one after the other. Some parts lost power for only twenty minutes, whereas others lost power for up to eight hours.[10] Over 3 million customers—and there are usually several people per "customer," which is synonymous with "electricity meter"—lost their power, and a total of 4,000 megawatts of capacity was shed. The after-action report that investigated the blackout episodes concluded that a vast preponderance of the failures were from frozen water in sensing lines, equipment, and water lines.[11] Who knew that little old pipes frozen with water could bring the whole grid crashing down?

In the wave of polar vortices that occurred in early 2014, there were a few additional instances of frozen water disturbing the energy system. Above Niagara Falls, concerns about ice damaging the hydroelectric power plant made headlines, causing authorities to bring out ice-breaking ships to protect the equipment and to allow the water to spin the turbines again.[12] Ice also prevented thermal power plants from working. During the same cold wave, ice on Lake Superior prevented coal boats from delivering fuel to power plants. The U.S. Coast Guard had heavy-duty icebreakers working overtime trying to clear lake ice, and ice cutters escorted the coal ships. Because of blocked deliveries, the Presque Isle power plant was running out of coal and bought 25,000 tons of excess stockpiled coal from a nearby power plant operated by Marquette Board of Light and Power.[13] Because there were no trains or barges between the two locations, the coal had to be transported by truck, which is an unorthodox and expensive way to move coal.

And, in one series of cascading failures that same winter, frozen water in the Midwest undermined the energy system, creating life-or-death situations.[14] Rivers and lakes froze, which prohibited the passage of barges carrying road salt. Because road salt couldn't get delivered, roads couldn't be deiced. Because the roads were icy, diesel trucks carrying propane had trouble delivering heating fuel to customers freezing to death in the upper Midwest. Making matters worse, earlier in the year, a relatively wet growing season yielded a bumper crop of corn for biofuels that was wetter than normal. Consequently, propane demand was very high as the fuel is used to dry the corn for long-term storage. In this case, excess water caused higher energy demand for drying corn, after which frozen water made it harder to deliver salt, which meant frozen roads inhibited deliveries of propane. So a water problem (icy rivers, too much rain for the crops) became an energy problem (propane shortages) exacerbated by another water problem (icy roads), which became an even bigger energy problem (disrupted energy delivery). The energy delivery system is more fragile than people think.

In addition to water temperature affecting the energy sector, the abundance of water also matters. In particular, water scarcity from extended drought can be a problem. Though hydroelectric power is attractive for many reasons, it is least reliable during droughts when the need to use water for other purposes such as drinking and irrigation might take precedence over hydroelectricity. For example, low water levels in hydroelectric reservoirs can force power plants to dial back or turn off.

Utilities reduce hydroelectricity generation when water scarcity occurs, and those reduced flows can also lead to higher rates for electricity. In some places, the risks are quite acute.[15] Although in the United States this phenomenon is an inconvenience, in other parts of the world, it's a tradeoff between power and quenching our thirst. For example, cities in

Uruguay were confronted with the choice during the 2008 drought of whether they wanted the water in their reservoir to be used for drinking or electricity. Ultimately, reduced hydroelectric power output can strain the entire grid in regions where hydro is a primary power source.[16] The Energy Information Administration's monthly update in April 2012 raised an alarm that the California grid would face a number of challenges in the summer, partly because of the lower-than-normal snowpack in the Sierra Nevada mountain range, which was expected to reduce hydroelectric output.

Returning to the example introduced earlier, in July 2012, the grid did come crashing down—in India—because of strains induced by decreased hydroelectric output.[17] This event was the largest blackout in history, affecting more than 600 million people. Drought, triggered by the later-than-normal monsoon season, played a role in two ways. First, there was reduced hydroelectric power, making it harder for the grid to meet demand. That problem was exacerbated by floods months earlier that caused runoff from farms to silt up behind dams, reducing their capacity.

Second, because of reduced rainfall, the need for electricity from farmers to pump water for irrigation was higher than normal, straining the grid. There were simply more farmers using pumps to move water, which drove up demand, causing different regions of India to overdraw compared with their allotted power. Because of a long history of overpumping combined with reduced recharge to aquifers, water tables have been falling for years in India. Farmers using groundwater for irrigation had to lift water over a higher elevation gain, which also increased energy demand. Furthermore, farmers using surface water had to pump the water greater distances to cover a greater swath of land, also increasing energy demand. All of those responses combined to cause spiking demand for power at the same time that hydroelectric generation was lower, pushing the system past its breaking point. Once the power was out, millions of pumps for wells used by farmers and households quit working, as did the water supply and wastewater treatment systems for municipalities. Thus, a water event (drought) caused an energy event (increased demand) that caused another energy event (power outage) that caused a water event (water supply shortages). Ultimately, the energy constraint became a water constraint.

The problem of water scarcity is not just limited to hydroelectric reservoirs: it also affects thermal plants that need vast amounts of cooling water. This problem was particularly acute during the extensive drought in 2007 and 2008 in the southeastern United States. Unlike the problems of heat waves and thermal pollution, which might cause a power plant to reduce its power output, if water levels drop below the physical location of water intake pipes, then the power plant ceases operating completely. During the drought in the southeastern United States, lake levels came within three and a half feet of the water intake pipe for the cooling system at the Harris reactor near Raleigh, North Carolina, and within one foot of the intake level needed for one of the backup systems at the McGuire plant near Charlotte. It is difficult to expand, lower, or lengthen the water intake pipes so that they are less exposed to falling lake levels. According to one AP report,

> Extending or lowering the intake pipes is not as simple at it sounds and wouldn't necessarily solve the problem. The pipes are usually made of concrete, can be up to 18 feet in diameter and can extend up to a mile. Modifications to the pipes and pump systems, and their required backups, can cost millions and take several months. If the changes are extensive, they require a [Nuclear Regulatory Commission] review that itself can take months or longer. Even if a quick extension were possible, the pipes can only go so low. It they are put too close to the bottom of a drought-shrunken lake or river, they can suck up sediment, fish and other debris that could clog the system.[18]

Power plants are amazing testaments to human ingenuity and innovation. They are also hard, slow, and expensive to adapt because they depend on large physical structures.

Lake Lanier, a key reservoir in northern Georgia that provides drinking, irrigation, and cooling water, reached its record low level in December

2007.[19] Consequently, power plants near Atlanta during the winter 2008 drought were within days of shutting off because the vast amounts of cooling water were at risk from diversion for other priorities such as municipal use for drinking water. At the same time, amazingly, there was no restriction on municipal lawn watering, which drained supplies from Lake Lanier even further. Overall, 24 of the United States' 104 nuclear reactors at the time were sited in the region that endured that drought.

When drought and heat waves are combined, as they often are, the problems are compounded. For the same heat wave in France in 2003 described earlier, there was a simultaneous drought that caused France to lose 20 percent of its hydropower capacity.[20] France gets 78 percent of its electricity from nuclear power and approximately 11 percent from hydropower, which made this drawback hard to accommodate. The same problems happen in the United States. In one instance in August 2007, the Tennessee Valley Authority shut down one of three reactors at its Browns Ferry nuclear plant because drought lowered water levels in the Tennessee River, which was its source of coolant. Low water levels with the hottest temperatures in several decades plus water warmed by upstream power plants heated the available water above acceptable thresholds. The same river's tributaries were also low, reducing the output from neighboring hydropower plants.

The drought lowered the water levels in the rivers, which made them more vulnerable to increased temperatures from the heat wave, causing a violation of the thermal pollution limits. The power problems were exacerbated by the fact that the same drought that caused problems for the thermal plants also reduced output from the hydroelectric power plants, causing upward pressure on power prices.

Drought causes other problems for the energy sector. Severe drought in 2012 threatened a two-hundred-mile stretch of the Mississippi River from St. Louis to Cairo, Illinois, with the risk of barge traffic being shut off in early 2013.[21] The Mississippi is one of the world's most important inland navigation routes, moving several billion dollars of goods each month. But because of the drought in 2012, the Army Corps of Engineers decided to save water for summer irrigation in its upstream reservoirs. Consequently, releases were reduced, diminishing flow to the Missouri River, which subsequently reduced the flow into the Mississippi River. If water levels get too low, barges heavily laden with goods would not be able to travel safely. Those barges carry things like fertilizers, salt (for winter road treatments—the same kind of salt that couldn't be delivered over frozen rivers in 2014, causing energy shortages in the upper Midwest), and agricultural products. They also carry a lot of coal to midwestern power plants. That means a water scarcity event can threaten the supply chain for the energy industry in yet another way. And, in this case, water for irrigation was deemed more critical than water for the power industry. It is hard to imagine shipping the tonnage of coal and refined fuels by truck instead of barge, just because of a limit to the availability of trucks and capacity of the road systems. Furthermore, doing so would surely drive up prices dramatically.

In addition to water being too hot, too cold, and too scarce, it can also be too abundant. It is hard to imagine that water abundance can be a problem, given the prior discussion about the risks that droughts pose to energy production. But too much water also puts power plants at risk of damage. In one case a Nebraska nuclear power plant nearly shut down because of flooding of the Missouri River in June 2011.[22] The floodwaters came within a few feet of cresting over temporary, inflatable flood walls surrounding the facility. One of the berms collapsed, triggering concerns of the need for a shutdown to prevent safety risks. And, for the Fukushima incident in Japan in 2011, it was not the earthquake that caused the biggest problems. Rather, the tsunami and the subsequent flooding ruined backup safety systems and triggered the meltdowns and explosions within the nuclear power plants. Those same floods can also overwhelm water and wastewater treatment systems, causing a spike in electricity demand.

The problems that water constraints cause for energy are not restricted to just the power sector: they also affect fuels production. The problems of water show up in a variety of ways. For extracting minerals such as coal, water constraints are not as prominent. Water use for washing and dust control does occur for coal, but it is relatively minor in terms of the overall production, and that means coal production is not threatened very severely by water shortages. Conventional oil and gas production use water-flooding in some reservoirs to enhance productivity, and so in those cases, water scarcity might restrict production.

Biofuels, because they require so much water for their growth, are the most sensitive to water constraints. Floods also destroy energy crops and cause massive loss of soil and runoff. Droughts hurt agricultural production in general, and energy crops such as corn and soy for ethanol and biodiesel are not spared. When the epic drought of 2012 hit the corn belt, corn production dropped and prices for ethanol increased. Also, some cities, including even Champaign, Illinois, in the middle of the corn belt, feel that they are in competition with agricultural users of water for energy crops.[23]

New techniques like hydraulic fracturing are also vulnerable to water constraints. Many of the sites with booming shale production, such as the Eagle Ford Shale in south Texas and the Bakken Shale in North Dakota, are in areas that are drought-prone. The shale producers may find themselves angering locals as a result. Because the oil and gas producers are often the newest local users of water compared to the entrenched users such as agriculture and municipalities, they arouse a lot of resentment when they suddenly show up for an oil and gas boom and purchase a lot of water to meet their production needs. Some towns are already banning the use of municipal water for fracking. Officials for the Ogallala Aquifer included hydraulic fracturing when they approved restrictions on water use in the groundwater district.[24]

At the same time, the shale producers are willing to pay a much higher price for the water than conventional users. They can afford to pay higher prices because the oil and gas that are produced are so valuable, and by contrast the price of water is very low. According to the CEO of Breitling Oil and Gas, Chris Faulkner, the company paid $68,000 for 3.5 million gallons of water, which is 0.2 percent of the $3.5 million spent to hydraulically fracture the well.[25] In total, water procurement, injection, collection, trucking, and disposal for fracturing operations can be as high as 10 percent of the total cost of an active drill pad.[26] It is safe to conclude that the availability of water, not the price, would be a key constraint.

While the risk to fuel production is usually whether there is enough water available at the front end to perform the extraction, the more important risk for shale production might have to do with the wastewater that is generated at the back end of production.[27] Shale gas wells produce significantly less wastewater per unit of gas that is ultimately recovered than conventional gas wells, but the abundance of wells in a small geographic area that has not historically had a lot of production means that the flows of wastewater can exceed the local capacity to deal with them. How effectively the wastewater from shale wells is managed will be the ultimate constraint on production.

The total wastewater from an unconventional well comprises three different flows. There are the fracturing fluids (or "frac fluids") that are injected into the well. These typically include freshwater along with a mix of chemicals designed to enhance productivity through a variety of means, for example by reducing friction, chemical precipitation, scaling, and biological fouling and changing viscosity. These chemicals raise public concern because of the potential risks they pose to ecosystems and human health. Of the millions of gallons of frac fluids that are injected into the wells, a major fraction comes back to the surface as "flowback" water.

In addition, significant volumes of "drilling fluids" or "drilling muds" are brought back to the surface. Drilling fluids are used to cool and lubricate the process of cutting through the rock while also clearing out the cuttings by bringing them to the surface. These drilling fluids have high concentrations of total dissolved and suspended solids, which complicates their disposal. At the same time, because horizontal drilling expands the length of wells compared to conventional drilling, it is reasonable to expect a lot more drilling wastewater.

Last, water is naturally present in the shale formations. When gas is extracted from the wells, this "produced" water comes with it.[28] Because the water is highly saline, with a density of total dissolved solids that can exceed 100,000 milligrams per liter or more, meaning it can be three times more salty than the ocean, it is sometimes referred to simply as "brine." But, it has many more components than just salt, including high concentrations of other materials that are of concern, such as metals, organics, arsenic, and sometimes naturally occurring radioactive materials (called NORM for short, though most people wouldn't consider them normal components of drinking water). While the "flowback" and "drilling fluids" are generated as wastewater during the drilling and fracturing phase of the well's life cycle, the "brine" continues to be produced over the well's lifetime.

In the Marcellus Shale, a typical well generated approximately 1.25 million gallons of wastewater, just over half of which was brine, one third was flowback, and the rest was drilling fluids.[29] The unconventional wells produce about ten times more wastewater than conventional gas wells, but they also produce about thirty times as much gas. So while the total wastewater volumes are higher for unconventional wells, which can create local environmental hotspots, the wastewater-to-gas ratios are actually more favorable. The unconventional wells have a gas-to-wastewater ratio about three times higher than the conventional wells over the first four years of production.

These are still high volumes of wastewater, and when multiplied by thousands of wells, the question becomes how to handle the waste. As the number of wells increase, so does the total wastewater flow. Initially, the wastewater is stored in surface ponds, which pose a risk to water quality if the ponds are not properly lined because the dirty wastewater can trickle down to the freshwater aquifer belowground. After that, the management options are limited.

There are a few ways to handle the wastewater: treatment, reuse for subsequent production, and disposal. Because of the high solids content of the wastewater, conventional municipal wastewater treatment facilities are not well equipped to handle the streams. Consequently, the dirty wastewater can pass through without being cleaned up sufficiently before it is discharged to the local rivers, which can elevate their pollution levels beyond acceptable thresholds.[30] Highly specialized industrial wastewater treatment facilities, which are more capable but also more expensive and energy intensive, are more appropriate. Unfortunately, the industrial facilities are rarely proximate to the drilling site, which means millions of gallons of wastewater must be transported by truck or pipe from the well to the facility. This movement of water is essentially an unintended interbasin transfer of water from one watershed to another. While that is not a big deal at small volumes, for many thousands or tens of thousands of wells, the volumes of interbasin transfer become nontrivial.

There is also the possibility for on-site treatment to reduce wastewater volumes. While this approach is an advancing technology, it is in limited use at this time. That approach, however, remains a critical technical solution to the problem, allowing producers to avoid trucking or piping their wastewater all over the state.

As discussed earlier, underground injection of wastewater is another method of disposal. While that method is common in Texas, it is uncommon in Pennsylvania (where it is essentially forbidden because of a lack of suitable injection sites). The disposal options for the Barnett and Eagle Ford Shales in Texas are therefore different than for the Marcellus Shale in Pennsylvania. Consequently, some of the wastewater from Pennsylvania was trucked over the state line to Ohio, where it is allowed to be injected. Unfortunately, deep injection disposal triggered a series of more than one hundred earthquakes in the span of just over a year, culminating in one on New Year's Eve 2011 in Youngstown, Ohio, that was 3.9 in magnitude on the Richter scale.[31] These temblors rattled Ohioans in a place where earthquakes are uncommon and building codes are not designed to meet seismic standards. What was particularly galling for many people in Ohio is that it felt as if Pennsylvania were reaping all the economic benefits of shale gas production while Ohio felt the brunt of its drawbacks.

It is also possible to dispose of the wastewater through "road spreading," which is basically just what it sounds like. Admittedly, that does not sound like a very appealing option and is prohibited for Marcellus wastewater, although it does have some useful purposes such as ice and dust control. Other reuses include recycling the wastewater to use in subsequent wells. In fact, that approach, while more expensive today than injection, is growing rapidly and might be an effective solution to the treatment and disposal constraints. This method is not without its problems, however: the wastewater has high solids concentrations, which can inhibit production from shale gas wells that are second or further down in line. The presence of bacteria in reused water can trigger biological phenomena such as fouling and the creation of corrosive materials, which also downgrade production. In places where water for frac fluids and wastewater disposal are expensive, recycling is likely to be an economic solution.

The key challenge is that the sudden growth in wastewater volumes in production areas can exceed local infrastructure and capacity to handle it. So, the shale gas wells, popping up by the thousands, overwhelm the local systems, ultimately constraining production on the back end of the life cycle. In the final analysis, it has been determined that the risks from fracking are more acute for water quality than water quantity.[32]

Just as water limitations can restrict energy, energy limitations can also restrict water. One of the key risks to the water system is its energy dependence: because the water sector requires so much energy, energy disruptions can cause water disruptions, especially for the water and wastewater treatment plants. This phenomenon is particularly true for power outages. Electricity is not the only form of energy consumed for water—natural gas and petroleum are also used for water heating, for example—but electricity is the form of energy that is used for water pumping. That means when the power goes out due to storms or intentional acts, water treatment plants cannot pump water to serve their distribution networks and wastewater facilities quit functioning. For example, Hurricane Ike knocked the power out along broad swaths of the Texas and Louisiana coast, causing sewage pipes to stop flowing, creating a public health risk.[33] The consequences can trigger severe public health impacts if tainted water makes its way through the distribution system or if sewage goes untreated and is discharged into waterways.

Usually water systems pump water into elevated storage tanks. That means after a power outage water can still be served to many customers by gravity. But those water tanks do not get refilled when the electric pumps quit working, meaning the water pipes will go empty eventually. While it is possible for the natural gas grid to fail—for example when demand is really high for residential heating—it is more likely that the electrical grid will be disrupted. The primary difference is that the gas grid is belowground and the electrical grid is mostly aboveground, making it vulnerable to high winds that knock the lines down or knock trees down into the lines.

The outages described earlier from Hurricane Katrina, which affected water supplies in New Orleans, make up another prominent example. The impact of power outages does not just have to be on a large scale: small-scale power outages can also be highly disruptive. For example, for houses that get their water from wells that operate with electric pumps, power outages become water outages.

I happen to live in a neighborhood in a suburban area of Austin that until very recently used well water to serve ninety-five homes, despite the fact that we are surrounded on all sides by conventional piped-water systems. In May 2011, a little less than a year before the small well with its electric pump was connected to the neighboring piped systems, it suffered a power outage. The well operators taped a notice to the front door at my house that warned us to boil our water prior to consumption "as a precautionary measure." Even in modern times in rich cities in urban, industrial areas, we have to periodically boil our water to purge it of health risks such as water-borne pathogens, all because of an energy constraint. In this case, bringing the nexus around full circle: we use energy to fix problems with water that were caused by energy outages.

In the end, boiling water is pretty easy for someone like me, as all I have to do is turn the knob on my stovetop. If I do not want to do that, I can drive to the nearby grocery store to get some bottled water to meet my needs until the water service is restored. However, in developing countries, boiling water is a time-intensive, laborious task. While the energy outage is a mere inconvenience for me, it is a women's rights or human rights issue in other parts of the world.

Though centralized power outages can knock out the water treatment system, distributed power outages in just a neighborhood or at an individual home don't stop the entire water supply. Many people might have experienced a power outage at their homes during a windstorm. Even though the power goes out, in many cases the phone lines still work. In fact, historically people have needed to phone the power company to let them know the power was out. And, in almost all cases, the water stays on because it is fed by gravity into the home.

Adjusting to a power outage is inconvenient, but it can usually be managed with candles providing light and batteries providing backup power for computers and other electronics. It is annoying, but does not cause severe discomfort. The bigger inconvenience by comparison is a water outage. The patience to go without flushing toilets, operating sinks, or taking showers is much harder to come by than the ability to go without electricity.

The energy outages also create public health consequences. For example, after Hurricane Sandy created large surges of stormwater and knocked out the power at wastewater treatment plants, hundreds of millions of gallons of stormwater mixed with untreated sewage passed through and spilled into waterways.[34] At least six treatment plants in the New York area alone shut down completely, and many others were impacted in some way or another. Because of the stormwater floods combined with the failed pumping systems, the wastewater treatment plants got backed up, causing the sewage to flow the wrong direction back out of the drainage pipes. In a scene reminiscent of the science fiction movie *The Blob*, where a mysterious biological blob attacks whole cities, the *New York Times* noted that in one neighborhood "a plume of feces and wastewater burst through the street like a geyser." Raw sewage was still seeping into homes at least five weeks after the hurricane struck. As the media coverage noted at the time, that consequence might be one of the longest lasting legacies of the hurricane. For these treatment plants, the electrical equipment was flooded, causing them to switch off. Making these treatment plants more resilient will require elevating the electrical equipment above the flood line and waterproofing the equipment. Only in this way can cities avoid a water event that triggered an energy event that triggered a water event.

SEVEN

Trends

THE KEY UNDERLYING DEMOGRAPHIC trend that can strain the energy-water nexus is growth: population growth and economic growth. Population is growing, exceeding 7 billion people globally in 2011, and is projected to continue growing. Peter Gleick, a leading scholar on water issues, MacArthur Fellow, and member of the National Academy of Sciences, made a comment about global population that really stuck with me. While briefing the Roundtable on Sustainability at the National Academy of Sciences meeting in June 2013, he discussed global population trends. He said, "The most interesting day in the history of the world will happen in the twenty-first century: that is the day the global population is smaller than it was the day before." Until that day, global population is projected to keep growing, plateauing between 9 and 11 billion people sometime between 2050 and 2100.

Along the way, each one of those billions of people will need energy and water. More people means more demand. At the same time we have been getting richer. Demand for energy and water have been growing faster than population, driven by economic growth on top of the population growth.[1] This phenomenon occurs because affluent people eat more meat, which leads to water consumption. They also consume more electricity, which uses water. Unfortunately, many water withdrawals are from nonrenewable resources. That means the trends for consumption will trigger water shortages unless something changes. By 2005, at least half of Saudi Arabia's fossil (nonrenewable) water reserves had been consumed in the previous two decades. Globally, much more groundwater is pumped out of aquifers than is recharged naturally. Therefore the water table has been going down.

It's not just the Middle East, though. Significant declines have also been observed in the Ogallala Aquifer under the Great Plains of the United States, spanning eight states from South Dakota to Texas.[2] Water tables lowered by as much as 234 feet were observed in Texas, while the average drop across the entire aquifer was 14 feet. Storage of water fell from 3.2 billion to 2.9 billion acre-feet. Although these numbers are daunting, the impacts on water in the aquifer were not exclusively negative in all locations. Some localized increases up to 84 feet were observed in Nebraska as a consequence of seepages and reservoirs affecting the amount of water stored underground.

These withdrawals from nonrenewable sources and shifts in water levels, stored water, and water use over time affect the amount of water that is available to humanity and nature. Overall, water availability is declining globally.[3] Available water dropped from 17,000 cubic meters per person in 1950 to 7,000 cubic meters per person in 2000. Water stress occurs between 1,000 and 1,700 cubic meters, and a water crisis occurs at less than 1,000 cubic meters. Notably, some countries are already at these levels, and are cause for concern: Qatar (91), Libya (111), Israel (389), and the UK (1,222). All of these datasets point toward a conclusion that water stress is increasing. High-profile research published in *Nature* has concluded that nearly 80 percent of the global population endures high levels of threat to water security.

Perhaps one of the most impressive signs of overpumping is that we are actually causing the oceans to rise in a nontrivial and measurable way.[4] Scientists have found that a variety of water-use factors have caused the sea level to rise 0.7 millimeters per year between 1961 and 2003, which is about 42 percent of the observed total sea-level rise. These factors include overuse of groundwater, artificial reservoir water impoundment, and climate-driven changes in terrestrial water storage. Of the 1.8 millimeters per year of total sea-level rise from 1963 to 2003, climate change caused 1.1 millimeters of the rise, through heating of the oceans, which caused thermal expansion, and melting snowcaps and glaciers, which caused additional runoff of freshwater into the oceans, raising sea levels. Land-based groundwater extraction with subsequent runoff was responsible for the other 0.7 millimeters, a substantial portion.

Climate change is likely to exacerbate strains at the energy-water nexus. The way climate change manifests itself is through changes to the hydrologic cycle. "Climate change" might be better named as "water change." And those changes show up in a variety of ways, such as elevated ocean levels, elevated ocean temperatures, more frequent and intensive flooding, more frequent and intensive droughts, and distorted snowmelt patterns.

These shifts can have a significant impact on civilization, because our societies have built themselves in particular locations and with specific configurations based on expectations built over centuries for where the water will be. Where and how much water is available at which time of year has been a driving force for the industrial mix, agricultural choices, and many other societal elements that we take for granted today. Most societies configured themselves with an expectation that water availability would stay the same, making themselves vulnerable to sudden, dramatic changes. These vulnerabilities mean that society can actually collapse in the face of extended changes to the availability of water.

Elevated ocean levels are a direct threat to the 40 percent of the world's population that lives within about sixty miles of the coastline. Higher ocean levels raise the risk for erosion of coastlines, submersion of valuable properties and infrastructure, and saltwater intrusion into freshwater aquifers. All of these are expensive to mitigate. Elevated ocean temperatures have ecosystem impacts that might be bad for fisheries, aquaculture, and power plant cooling. More frequent and intensive flooding is difficult for societies to manage for obvious reasons. Floods are hard to control and can do a lot of damage. More floods mean more cumulative damage, and greater intensity means each individual flood is likely to be more damaging or sudden than usual. It will cost a significant amount of money to move buildings out of expanded flood plains or shore up levees, protect land that can absorb the water, or build reservoirs that are intended to stay empty and only used for capturing excess water during flooding.

On the flipside, more frequent and intensive droughts will be the ironic partner to the floods. Mitigating droughts requires expensive infrastructure for storing water, long-haul pipelines to move the water farther, more powerful pumps for raising water from ever deeper wells as surface water sources dry up and overextraction from nonrenewable groundwater sources increases.

Distorted snowmelt patterns will be another consequence of climate change. The snowpack may be thinner and melt earlier, affecting the rhythms of water availability, irrigation, crop rotation, and other patterns that have built up over centuries. Of the world's 7 billion people, approximately 1.5 billion rely on snowmelt from the Himalayas alone. Add in those living on snowmelt from the Rockies, Andes, and other major mountain ranges, and the tally will grow. The villages around Kilimanjaro have been flagged as particularly vulnerable. On May 29, 2015, the snowpack in California was officially reported to be zero, meaning the source of water for tens of millions of people and the vast preponderance of national fruit, nut, and vegetable cultivation was at risk.[5] Managing these shifting patterns might spawn impactful, expensive, and energy-intensive investments in large-scale water storage infrastructure such as reservoirs to hold the water over a greater span of time.

All of these outcomes can be mitigated in one way or another, either through investments in new infrastructure, changing industrial and agricultural mixes of the societies involved, or by picking up and moving to another location that will have better odds in the climate change sweepstakes. All of these options represent hard choices. And some of those choices, because of their energy requirements, might exacerbate the situation in the long term. At the same time, these options often fall hardest on the poorest societies. That means the emissions from the richest members of the globe will cause expensive problems for the poorest. The moral challenge of this situation is difficult to swallow. The inequality in the emissions (mostly by the rich), and the suffering (mostly by the poor) presents a key quandary for the world to resolve.

Unfortunately, the energy-water-climate nexus has a positive feedback loop. Our energy consumption causes climate change, which changes the

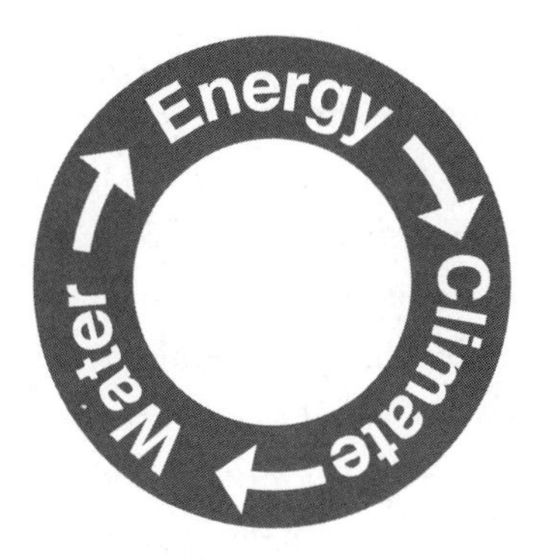

Energy, water, and climate have a positive feedback loop. Our energy sector contributes to anthropogenic climate change, which changes the water cycle, which causes us to use more energy to solve our water problems, which contributes to climate change, and so forth. [Image idea courtesy of Jane Long]

hydrologic cycle, triggering investments in energy-intensive water solutions, which exacerbates climate change, and so forth. Frustratingly, the higher temperatures of a warming planet reduce the global photosynthetic efficiency. That means we will use more energy-intensive irrigation, fertilizing, and harvesting with energy inputs to overcome the hit on efficiency.

Climate change may also reduce the amount of energy we get from emissions-free hydropower. In the United States, more than half of the nation's hydroelectric generation occurs in California, Oregon, and Washington. This region is also particularly sensitive to climate change: as the climate warms up, the snowmelt and precipitation patterns become distorted in ways that are detrimental. And, the cumulative impact of the changes is nonlinear and magnified. For a large basin like that of the Colorado River, small declines in precipitation cause major droughts, which in turn can dramatically reduce power output from a whole chain of hydroelectric dams. Every 1 percent decrease in precipitation causes a 2–3 percent drop in streamflow, and every 1 percent decrease in streamflow in the Colorado River Basin yields a 3 percent drop in power generation.[6] At the same time, many millions of people depend on that basin's water for irrigation, drinking, commercial activity, industrial processes, and, of course, for power production. And the outlook for precipitation may get worse. Higher temperatures also mean that there will be additional evaporation, reducing water stored in reservoirs. The reduced hydropower in California during the multiyear drought from 2011 to 2015 caused electric rates to increase: as hydropower dropped from 18 percent to 12 percent of the fuel mix, utilities spent extra money purchasing natural gas to make up the difference.[7] There is a silver lining, which is that hydropower could initially increase because of higher-than-normal snowmelt.[8]

Beyond the trends for increasing demand for total consumption of water and energy, we are moving toward more water-intensive energy. That trend is especially true for transportation fuels, where for national security, environmental, and economic reasons there is a strong desire in the United States to move away from petroleum. The preferred alternatives are domestic, low-carbon, and sustainable fuels. For many people, especially the agricultural sector, that means corn-based ethanol. But it could also mean natural gas, methanol, or electricity.

The challenge is that many of those fuels are more water intensive than conventional petroleum-based fuels such as gasoline and diesel.[9] Because biofuels require so much water, the federal push for more biofuels with the RFS (Renewable Fuels Standard, which requires that a certain volume of biofuels are consumed annually) has essentially become a mandated increase in water consumption for transportation fuels. The push for electric vehicles also has the unintended consequence of increased water use for power plant cooling. The RFS and incentives for electric vehicles are classic examples of energy policymaking on one hand that ignores the water consequences on the other hand. Adding up the biofuels volumes that are mandated will cause significant increases in water needs. In 2005, petroleum-based gasoline required about 250 billion gallons of water to produce 140 billion gallons of fuel. Switching to ethanol from corn—with just 15 percent of the crop requiring irrigation—means we will need well over a trillion gallons of water per year within two decades.[10] Just a small irrigated fraction of the biofuels mandate will cause water consumption for light-duty transportation fuels to go up by a factor of four or more. Just imagine how bad it would be if all the corn we grow required irrigation.

Adding in the expectations for other fuels such as cellulosic ethanol, coal-to-liquids, and other sources adds in yet another trillion gallons of water consumption. Keeping in mind that the nominal annual water consumption in the United States is about 36 trillion gallons, this 2-plus trillion gallons per year of additional water consumption is significant. It moves transportation into a category as one of the largest water consumers in the nation. As a nation we prefer enriching midwestern farmers instead of Middle East autocrats, which is an admirable goal. However, the water impacts of doing so with the RFS are significant; essentially we are switching from foreign oil to domestic water. Before embarking on such an ambitious mission, maybe we should check first to make sure we have the water. The story is similar in other parts of the world that are trying to displace conventional petroleum with thirstier options.

Similar to the trend of moving toward more water-intensive energy, we are also moving toward more energy-intensive water. This shift has several different components, including stricter water/wastewater treatment standards, deeper aquifer production, long-haul pipelines, and desalination. Each of those elements is more energy intensive than conventional piped water today, and seems to be a more common option moving forward. The market trend for bottled water could also be considered one of those energy-intensive options.

As societies become wealthier, their concerns shift from focusing on economic growth to protecting the environment. This phenomenon is described by movement along the environmental Kuznets curve. In the United States, we went through a similar trajectory. The first hundred years after the second Industrial Revolution saw significant increases in energy consumption. Then, since the 1960s, environmental protections have become more important, yielding several pieces of landmark environmental legislation in the early 1970s: the Clean Water Act, Clean Air Act, Endangered Species Act, and creation of the Environmental Protection Agency.

Many other prominent pieces of environmental rules have since been implemented.

Protecting drinking water quality from the output of water treatment plants for the sake of public health and discharge water quality from wastewater treatment plants for the sake of ecosystems are two important pieces of that trend. But water and wastewater treatment require nontrivial amounts of energy. Furthermore, advanced treatment methods to meet stricter standards are more energy intensive than treatment for lower standards. For example, advanced treatment systems for wastewater with nitrification require about twice as much energy as trickling filter systems. As we tighten the standards for water and wastewater treatment, we are essentially edging toward increases in energy consumption. While new treatment technologies and methods become more efficient over time after their initial implementation, the standards tighten in parallel. How these balance out is unclear.

At the same time, the water coming into water and wastewater treatment plants is getting more polluted with time. As population grows, there are more discharges into the waterways. Those discharges contain constituents that weren't always there in such high concentrations. For example, there have been growing concerns about pharmaceuticals (including birth control pills and pain pills) in sewage streams, which are difficult for wastewater treatment plants to remove. Doing so requires new equipment and ongoing investments of energy.[11]

In an ironic example of the energy-water nexus, some of our energy choices create water quality impacts that require additional energy to treat. For example, increased biofuels production from corn in the middle part of the United States is expected to cause additional runoff of nitrogen-based fertilizers and other pollution. Subsequently, we will need more energy to remove that pollution.[12] And many domestic users of water rely on their own personal wells to access untreated, clean, groundwater. If pollution infiltrates the groundwater, as has happened in the corn belt, users might need to add treatment systems, increasing their energy bills for their water.

The wastewater streams from hydraulic fracturing of shales to produce oil and gas contain much higher levels of total dissolved solids than most wastewater treatment plants can handle while complying with discharge standards.[13] That means more energy has to be spent in one of several ways: on trucking that wastewater to disposal sites or specialized industrial wastewater treatment facilities that might be far away (something that happens rarely), for on-site treatment to recycle and reuse the water in subsequent wells, or on new equipment at the wastewater treatment plant to treat those streams. Even that new equipment is sure to require energy.

We are also contemplating moving water farther from its source to end-use. Long-haul pipelines and interbasin transfer, which is moving water from one river basin to another, are common proposals to solve the crisis of declining local water supplies. While the idea of aqueducts has been around for thousands of years, the scale, length, and volumes of water that are moved are growing. Some of the classic water transfer systems include the State Water Project in California described earlier. California's system is the state's largest electricity user because it must pump the water over mountains, though it also captures a lot of energy when the water flows back downhill through inline hydroelectric turbines coupled with chutes.

In addition to the California system, the island of Maui has an incredible handcut series of water channels that circle its two volcanoes, moving water miles from the wet portion of the island—one of the wettest places in the United States—to the dry inland plains where farming occurs. This system operates by gravity, and also generates electricity along the way.

Moving forward, as water tables fall and surface sources dry up, municipalities are more likely to consider the cost of expensive and far-flung water-gathering systems that pull water to a city from deeper in the ground or farther away. These long-haul systems will generally not be gravity-fed, and will require a lot of energy. Plus, they will incur ecosystem impacts as water from one basin is moved to another. While that might be good in terms of supplementing the flow for the receiving basin, it is bad for the basin that loses the flow. Plus, there is the concern of invasive species moving along with the water.

Perhaps the most ambitious project is the South–North Transfer Project in China (also known as the South–North Water Diversion Project, or SNWD). This project essentially aims to move major southern rivers—the

Yangtze and Han—across the country to the Yellow and Hai Rivers. The industrialized north is relatively water poor, whereas the southern part of China is relatively water rich. The scale, scope, and ambition of the project is reminiscent of U.S. water planners who have dreamed for decades of diverting the Yukon River in Alaska or the Missouri River to the Southwest so that the deserts would bloom with flowers and fruit trees. Peter Gleick refers to these as "zombie water projects" because no matter how expensive or silly, the ideas just will not die.[14]

The total estimated flow for the Chinese endeavor is projected to divert 44.8 billion cubic meters per year from the south more than a thousand miles to the north, at a total cost estimated to be $62 billion.[15] Not to be left out, India is also building its own long-haul water pipeline. And, joining the pack, Texas is, too.[16] For example, in Texas, a 240-mile pipeline is being built to bring 370,000 cubic meters per day of water from Lake Palestine to the Dallas–Fort Worth metroplex. The total capital cost for the construction is estimated to be $888 million, or $3.7 million per mile of pipeline. The annual electricity consumption is expected to be $11.3 million, or $0.71 per cubic meter.

In addition, there is a water pipeline that oil and gas tycoon T. Boone Pickens proposed in early 2008 with the expectation that water would be the new oil.[17] The pipeline would move water from Roberts County in the panhandle of Texas toward the Dallas–Fort Worth metroplex. This project was controversial for a variety of reasons, one of which is that the water rights Pickens holds are for fossil water in the Ogallala aquifer, which can take millions of years to recharge. And by sending it to Dallas, it seems one of its likely applications will be for watering lawns. While some energy would be used for pumping the water out of the aquifer, once it is at the surface, it would mostly use gravity for its downhill trip to Dallas. Ultimately the deal was scuttled because of the $3 billion price tag for the pipeline. Instead, Pickens sold the water rights to local thirsty cities.[18]

Another of the key trends to watch is how many municipalities are turning to desalination as a solution for water supply issues. Some people have called this "a river flowing back from the sea." Traditionally, rivers flow from freshwater sources to saltwater destinations, picking up salts and other minerals along the way. In fact, the oceans were once fresh, but became salty over many eons of runoff from lands. Desalination reverses this trend, as the saltwater sources become freshwater at their destination.

In 2013, over 17,000 desalination plants were already installed worldwide, providing approximately 21 billion gallons per day (67 million cubic meters per day) of freshwater.[19] With a blistering pace of growth, that capacity is projected to keep expanding quickly. More than three-fourths of new capacity will be for desalinating seawater, with the rest from brackish groundwater or salty rivers. While thermal desalination (using heat) represents about 25 percent of the installed capacity by 2010, it represents a shrinking share of new installations as builders seek the less energy-intensive reverse osmosis membrane-based system. Even with the lower energy approach, desalination is still an order of magnitude more energy intensive than traditional freshwater treatment and distribution. Desalination is capital intensive, too: the annual global desalination market exceeds $10 billion.

Growth is particularly rapid in energy-rich, water-poor parts of the world, such as the Middle East, northern Africa, and Australia. After a severe drought that lasted several years, water-strapped Israel famously turned to the sea for its water, rapidly building a handful of desalination plants to produce about 200 billion gallons of freshwater annually by desalting water from the Mediterranean.[20] Rapid growth is also occurring in China, where booming industrial activity is straining water supplies that serve the world's largest population. It is also popping up in locations such as London and the United States, where the abundance of water is very different than in the arid regions of the world. As noted earlier, London's desalination plant was very controversial, and became a big part of several mayoral campaigns.

Despite its relative water wealth, the United States is the world's second-largest market for desalination, trailing only Saudi Arabia.[21] This phenomenon is partly the result of the unequal distribution of water resources across the United States. And, as a wealthy country, the water consumption per capita is quite high and the money to finance large-scale infrastructure projects is available. Projects are under consideration for seawater reverse osmosis in coastal states such as California, Texas, and Florida. And brackish water projects are under development to serve inland communities that sit atop large brackish aquifers, as in Texas, Arizona, and New Mexico.

The two most energy-intensive options—desalination and long-haul transfer—can also be combined to create an even larger energy requirement for water.[22] Natural water flows occur by gravity, but for seawater desalination, the opposite is true. By definition, coastal waters are at sea level, so moving the water inland requires pumping water uphill. One such desalination project under development in the United States is a coastal facility along the Gulf of Mexico that is designed to provide freshwater for San Antonio, Texas. That means the water would be moved nearly 150 miles inland, increasing in elevation nearly 775 feet.

While trading energy for water makes a lot of sense in places like the Middle East or Libya, where there is an abundance of energy and a scarcity of freshwater, that tradeoff is not obviously a good value in places like the United Kingdom or the United States, where other cost-effective options such as water conservation, graywater capture, and water reuse might be available.

EIGHT

Technical Solutions

Necessity is the mother of invention.

—Plato, *The Republic*

WHILE THE TRENDS TOWARD MORE energy-intensive water and more water-intensive energy carry a sense of foreboding, thankfully there are several different technical solutions that exist. These options include source switching, advanced technologies that reduce resource intensity, smart technologies that increase information intensity, distributed technologies, and integration of our water and energy systems so that one benefits the other. In some cases these solutions are already being implemented somewhere worldwide, and just need broader adoption. In other cases, much innovation remains to be done. However, all of these are plausible and technically feasible.

Some of the solutions have catchy slogans. "More crop per drop" for efficient water use in agriculture, "showers to flowers" for using graywater in our homes to irrigate our lawns and gardens, and "toilet to tap" for turning reclaimed wastewater effluent into drinking water. For whatever reason, similar slogans for the energy world have not been developed. Regardless, these and other approaches are worth pursuing.

Source switching means choosing energy sources with reduced water intensity or water sources with reduced energy intensity. Fuel switching can be accomplished in a variety of ways and has been done with a variety of motivations. In the 1970s, in response to the oil shocks, rampant fuel switching took place nationwide, to get away from the volatility and supply risks of petroleum and move toward other options. In the power sector, utilities switched from oil, which at its peak was responsible for 17 percent of our electricity generation, to natural gas, coal, and nuclear. By 2015, petroleum was below 1 percent of the fuel mix for the power sector. The industrial sector switched from petroleum to natural gas for process heat. And the residential sector, which had switched from coal to heating oil only a few decades before, started switching from heating oil to natural gas and electric heat. Many homes in the northeastern United States still use heating oil today, but the transition continues.

These shifts in the energy mix occurred for national security or economic reasons, but they also have environmental implications as we switch from dirtier to cleaner fuels or the other way around. Shifts also have water impacts. Consequently, we could intentionally shift our fuel mix to one that has lower water requirements.

In the power sector, many power plants built in the 1970s and 1980s are due for retirement or retrofit. These power plants are overwhelmingly coal and nuclear, built at a time when concerns about water scarcity were muted. In 1970, there were only 200 million people in the United States, compared with well over 300 million today. At the same time, that was a relatively wet decade. Because there was more water being used by fewer people, the power sector was not designed with water efficiency in mind. Planners at the time did not properly forecast the water strains that would occur from population growth, economic growth, and climate change.

Many of those power plants are very water intensive, with thirsty open-loop cooling systems designed around the steam cycle that was developed in the late 1800s. Since most power plants are designed and built for a nominal forty-year lifespan, many of those are coming to the end of their useful operation. To keep them going, substantial investment will be required, so most power plant operators are at a point where they must decide whether to retire or retrofit their power plants.

Nuclear power is very water intensive, so switching away from it would be a benefit from a water perspective. However, nuclear power plants are likely to have their operational lifetimes extended by at least twenty years for a variety of reasons. First, it is difficult to build new nuclear plants, and so extending the licenses of existing plants is usually a more straightforward path. Furthermore, already-built nuclear power systems are usually cost competitive, whereas new ones might not be because of their high capital cost for construction. Last, nuclear power plants, which do not have smokestacks, are well positioned to remain competitive as environmental standards tighten for emissions.

The story is different for coal. In addition to being very thirsty, many older coal plants are also very dirty: they emit high volumes of carbon dioxide, particulate matter, and mercury, which makes them vulnerable to more stringent emissions rules. Thus, the older coal plants are ripe for retirement. Phasing out those old coal plants and switching to natural gas combined cycle, solar panel systems, and wind farms all would spare significant volumes of water. Even when considering that natural gas produced by hydraulic fracturing from shale formations needs more water than coal mining, the reduced water use at the power plant because of the increased efficiency, avoided emissions controls, and partial air cooling for the combined cycle make the switch from old coal to new natural gas combined cycle plants a significant water saver overall.[1] The same approach is also relevant for the transportation sector: switching from petroleum to natural gas can save water, as could switching first-generation biofuels made from corn back either to petroleum or to biofuels that do not require irrigation such as sugar cane or cellulosic sources such as switchgrass.

Just as switching the energy source can save water, changing the water source can save energy or freshwater or potentially both. In particular, reusing water can be an appealing option. There is an Arabian proverb: "In the desert, any water will do." That includes degraded water: brackish water, saline water, graywater, and treated effluent.

If homes are built with distinct plumbing systems that separate the water streams, then the domestic graywater from showers, sinks, and clothes washers can be used within the home. Graywater can flush toilets inside the home or water our flowerbeds outside. Since it is not necessary to use the most energy-intensive form of treated drinking water for these applications, graywater usually represents an energy savings for nonpotable applications. In many ways it is ridiculous that we use the world's cleanest water for toilet flushing in the first place, so this approach seems sensible by comparison. With minimal treatment to remove the organic components, it can also be used again for washing. While graywater reuse can help avoid using the most energy-intensive water, if everyone does it at a large scale, then the sewers might not have enough water to operate effectively. These kinds of unintended consequences should be planned for. In addition, rainwater can be collected from storm runoff through roof systems of gutters into rain barrels. That water can then be used for irrigation. Notably, while rainwater might be very clean when it lands on the roof, it subsequently might pick up chemicals that make it unsafe to drink.

It is also important to note that if you have a graywater system at your home, then it would be good to use biodegradable soaps and avoid putting toxics down your drains, lest they end up on your plants a few hours later. The International Plumbing Code, which is only used in certain jurisdictions in the United States, allows for graywater from showers and bathtubs to be used for flushing toilets. However, the Uniform Plumbing Code, which is used more widely in the United States, prohibits graywater use indoors.

Harvesting graywater is not only a way to supply more water, it is also an option for harvesting heat. While the graywater from toilets generally is not heated—though some fancy Japanese toilets have heated water for user comfort—the water from sinks, showers, and laundry machines is often still hot when it goes down the drain. That heat is typically lost through the pipes to the ground as the wastewater moves along. However, with a heat recovery device, it can be used to preheat incoming freshwater to the water heater tank, saving energy.

Overall, graywater use can spare some freshwater withdrawals and might save energy for heating, pumping, and treatment. For some rural settings where water is really scarce, graywater might also be contemplated as a source of potable water. In these cases, graywater reuse for potable needs might actually be more energy intensive than getting freshwater straight from a well.[2] Many rural areas get water from wells, especially for remote domestic purposes or many small farms or ranches. For these applications, the groundwater is often very clean, and so the energy requirements are simply for pumping. By contrast, graywater reuse for potable purposes would require additional treatment to remove the organic components and pathogens, driving its energy intensity higher than simple pumping. In those cases it makes more sense to use the graywater only for irrigation. However, in municipal areas that already have advanced water treatment systems in place, graywater reuse could make a lot of sense.

One of the approaches to solve this crisis is to put the waste streams from the energy and water systems to useful purpose. There are a variety of ways to accomplish this goal. In particular, there are 30–40 billion gallons of treated effluent generated each day in the United States.[3] Because the wastewater is generated in cities, the wastewater plants are nearby. That means the effluent is an abundant source of water that is generally co-located with population centers. And, it is typically overlooked as a source. If that treated effluent is "reclaimed"—or used again—it can be a reliable supply of water. Reclaimed water is distributed in purple piping systems so that it can be distinguished from the treated drinking water and sewage.

Effluent is usually returned to lakes or rivers, or ejected into aquifers or oceans. While our ecosystems depend on many of those returns, it is also possible to use the effluent again before returning it to the watershed—for example, running the effluent through a power plant as a source of coolant. If the water does not need to be returned to a nearby river then it could be used for consumptive purposes as an alternative to freshwater—for example, for irrigating crops, hydraulic fracturing, or industrial purposes. The "toilet to tap" idea introduced earlier is a robust one. That approach saves energy compared with desalination to make brackish water or seawater potable. As it is already in place around the world in places like Singapore, Israel, and Southern California, and out of this world aboard the International Space Station, it is a proven technology.

For many municipalities, closing the loop with their waste streams to turn the wastewater into drinking water or for other water purposes might make a lot more sense from an economic or energetic perspective. Austin, Texas, has set up an extensive purple-piping network that brings the treated effluent from the wastewater treatment plant to downtown Austin, some high-density neighborhoods near the urban core, and the University of Texas, where it is available for irrigation and cooling. This approach is sensible since drinking water is not needed for these applications and because treated effluent in Austin is much cheaper to buy than drinking water. The original customers for the effluent were local golf

Treating wastewater effluent to a standard such that it is suitable for drinking is sometimes called "toilet to tap," which should not be confused with actually drinking water out of the toilet. It is not clear why a warning sign is necessary: who would drink toilet water? [Photo by Ashlynn S. Stillwell]

courses that used the water for their extensive irrigation requirements. Since then, more customers have emerged for the water. The effluent can be used within buildings for nonpotable purposes such as toilet flushing. In this case rather than "toilet-to-tap" it would be "toilet-to-toilet." For organizations that have the up-front capital to invest in the purple-piping infrastructure such as the city of Austin or the University of Texas, reusing effluent is a cost-effective option.[4] Generally speaking, reuse projects are more expensive than water conservation programs, yet more affordable than seawater desalination.[5] It is also a resilient, drought-resistant source of water supply. However, despite those benefits, it does pose some financial, health, and performance risks. Just one risk is the possibility that the purple-piping system will accidentally get cross-connected with the drinking water system, meaning treated effluent that isn't intended for drinking will be used for potable applications. And the billing rates for reclaimed water are typically set at a level below what is needed to recover the full costs of building the system, which means the customers are subsidized. While those customers surely benefit from the lower costs, the discounted rates put the long-term financial health of the reclaimed water system at risk.[6]

One place where effluent can displace the use of freshwater is energy production. In particular, hydraulic fracturing is a rapidly growing technique that uses water to improve productivity from wells in many more locations than before. In some areas, those water needs are competing with other users. At the same time, hydraulic fracturing does not require drinking water or even freshwater, for that matter. Effluent might be a useful alternative, and its use has already been demonstrated by some shale producers in Texas, where water scarcity encourages experimentation with water alternatives.

Effluent can also be used for power plant cooling.[7] In 2010, there were already forty-six power plants out of more than a thousand power plants in the United States that used reclaimed water for cooling tower makeup. Another handful of facilities use reclaimed water for cooling ponds, air scrubbers, as injected pressure at geothermal fields, and as boiler feedwater. Those include the very large nuclear power plants at Palo Verde in Arizona, which use the municipal wastewater from Phoenix and other cities nearby. While reclaimed water was first used for power plant cooling in 1967 in Burbank, California, in total only about a dozen of these cooling systems were built before 1990. After that, the pace of using effluent for cooling water accelerated, reflecting the increasing pressures on water resources in general.

Generally, effluent is safe to use for power plant cooling, although some minimum secondary treatments are recommended for disinfection and to prevent problems with equipment. For example, additional treatment can be useful as a way to prevent scaling, corrosion, fouling, foaming, or biological growth, all of which can degrade performance. In addition, there are concerns that windblown spray from the use of reclaimed water for cooling might reach workers or the general public, a risk which can be mitigated by setting the cooling tower at least ninety meters from the public or using higher levels of disinfection.

A general rule of thumb is that it makes sense to ship reclaimed water up to twenty-five miles for power plant cooling, beyond which the pumping requirements outweigh the benefits. In places where water is scarcer, then the twenty-five-mile rule is not as relevant. For example, the Palo Verde plants, located in a desert, must pipe the reclaimed water dozens of miles from the treatment facility to the power plant. The cities of Phoenix, Glendale, Scottsdale, Tempe, Mesa, and Tolleson all provide the wastewater for the power plant.[8] The wastewater from Phoenix moves nearly thirty miles downhill using the force of gravity, then is pumped uphill for eight miles to the nuclear facility. The other wastewater streams move even farther, up to sixty miles in some cases. Although that seems like a long distance, it is much shorter than the hundreds of miles that freshwater is piped from the Colorado River. Palo Verde is the only nuclear facility that uses 100 percent reclaimed water for its cooling. It is also the only nuclear facility whose cooling systems do not return water to the environment. The water reclamation facility can handle a total flow of 90 million gallons per day, and uses 25 billion gallons annually. And, while the power plant operators are willing to pay a high price and would like more wastewater, they now face competition from cities and other industrial users, who also want that water.

Scaling up the idea at other plants, the tens of billions of gallons of wastewater generated each day across the United States are more than enough to satisfy the billions of gallons of water consumed each day at power plants. The question of practicality becomes one of distance, cost, and other tradeoffs. While the total flow of treated wastewater is sufficient to meet the total consumption needs of power plants, there could be a spatial or temporal mismatch that prevents its convenient use. For example, large wastewater treatment plants and large power plants might not be close to each other. Or, the times of day or year when wastewater flows are highest may not correspond with the times of day or year when water is needed the most by power plants. And, many regions rely on wastewater discharges to streams, lakes, or aquifers. Diverting those discharges for power plant use instead might deprive those ecosystems of badly needed water. Despite these obstacles, it is clear that using reclaimed water for power plant cooling remains mostly an untapped opportunity. While only a few dozen plants use effluent today, there are hundreds that could.

Effluent can also be used for nonpotable industrial processes, such as firefighting, or as a heat transfer fluid in the form of steam for distributing heat or chilled water for cooling. And water is a necessary ingredient for snowmaking for ski resorts. Larger ski resorts often use snowmaking to increase the skiable terrain or to extend their operable season. A posh ski resort in Killington, Vermont, that caters to Bostonians and New Yorkers looking for a large ski hill within a half-day's drive of their city, uses its snowmaking as a key selling point. The snowmaking system uses more than 720,000 gallons of water per hour at full force, operates 1,500 snow guns with eighty-eight miles of pipe, and can add one foot of snow to eighty acres over the course of an hour. Killington uses effluent as its water source for snowmaking, noting in its promotional materials that it has "a virtually endless supply of water."[9] As wastewater is continuously generated, this claim is true.

And while effluent has its advantages in supply and cost, its use can generate criticism. When Killington first announced this plan in 1987, detractors poking fun at Killington created a fake organization named "Vermont Association for Sanitary Skiing" to generate opposition; they also created the bumper sticker "Killington: where the affluent meet the effluent."[10] This clever slogan is reminiscent of the "toilet to tap" phrasing in terms of its intent to generate a negative impression. And it worked: several Vermonters whom I befriended while writing this book over the river at Dartmouth College said they would never ski at Killington for precisely this reason. However, it turns out that Killington was simply decades ahead of its time in recognizing that properly managing a scarce resource is a sensible thing to do. Using effluent also offered cost savings compared with building reservoirs or using municipal drinking water. The snowmaking was good for business because it helped improve skiing conditions and allowed the resort to extend its ski season, both of which increased revenues. This case is yet another example of how thoughtful management of resources can carry economic and environmental benefits simultaneously.

Interestingly enough, one of the initial critics—Bill Mares, who was one half of a duo of legislators who were poking fun with their bogus organization and clever slogan—issued a public apology in 2012 for the gimmick. In an interview, he noted that he and his legislator compatriot found the whole episode amusing, but twenty-five years later in the context of heightened awareness of water scarcity, he admitted, "Now the joke is on me. Killington was way ahead of its time."[11]

Demonstrating the point, the idea found traction elsewhere. Arizona Snowbowl, a ski resort in the desert Southwest, announced its plans in early 2012 to "become the first ski resort in the world to use 100 percent sewage effluent to make artificial snow."[12] This situation raised protests because of health concerns. For example, what happens when skiers fall and accidentally ingest the snow? And, compounding difficulties, the mountain is sacred to the local Navajo tribe; spreading sewage on the mountain is equivalent to desecration.

Just as effluent or reclaimed water can be used to save energy and spare freshwater resources, the same can be done by using brackish or saline water as alternatives. Brackish or saline water can be used as drought-resistant sources of drinking water. Some of these sources are not obvious. For example, mine pool water from underground coal mines could be used—and in some cases is already used—by some power plants near coal mines in Pennsylvania and West Virginia.[13]

There are more than thirteen hundred power plants in the United States, about a thousand of which report their water use data to the U.S. government.[14] Of those, about 10 percent use either brackish or saline water. In total, forty-nine use brackish water (six from aquifers, forty-three from surface sources), and forty-seven use saline (two from groundwater, forty-three from surface water, and the remainder from "other" or plant discharge). Most of these power plants that use brackish or saline surface sources are coastal power plants that can use seawater. Seawater has the advantage of generally being cooler and abundant. Groundwater sources means the water is being pulled from brackish or saline aquifers, though it is unclear whether any water is being returned to the aquifers by the power plants.

As with effluent, there are some challenges with brackish and saline water for its users. For example, they are more corrosive than freshwater, which means better design and more expensive materials might be required. Sometimes demineralization helps, after which salt disposal is a burden. Also, coastal saline water is part of an ecosystem, so biofouling is important to monitor. Overall, the use of brackish or saline water for power plant cooling spares a lot of freshwater. The same could be said for hydraulic fracturing. While producers prefer to use freshwater to which an assortment of chemicals are added, there are some reports that brackish and saline water can also be used effectively.

For drinking water, desalination remains an option but is still limited. Solving its problems will help make it a robust option for societies. It is still expensive and energy intensive, and its brine stream is difficult to dispose of. But, with good engineering designs and advanced technologies, these problems can be overcome. One approach is to use integrated design for efficient thermal desalination, by which the waste heat from power plants starts the treatment.

Desalination can also be integrated with wind and solar to mitigate its energy and carbon footprint. One of the challenges with wind and solar is that their output varies for different meteorological and astronomical conditions. The position of the sun in the sky, the prevailing wind, and cloud cover all affect wind farms and solar panels. This intermittency is a challenge for grid operators, but less so if intermittent wind sources can be matched with an intermittent load such as desalination. Water does not need to be treated in a perfectly continuous fashion, so the water treatment facility can be dialed up and down to match the availability of wind. The same idea can be used for sunlight. This approach makes particular sense in places like west Texas where sunshine is abundant, where it is very windy, and where there is practically an underground ocean of brackish groundwater. Using renewables together in this way to treat the water solves several problems at once.

Biological approaches may provide other useful models. For example, mangroves grow in seawater, producing freshwater with pressure-driven ultrafiltration.[15] Enthusiasts have suggested that there might be a way to create mangrove plantations inside greenhouses or other capture systems, so that they could naturally produce freshwater at scale, without requiring massive input from fossil fuels. While this idea is appealing, it is akin to the idea of space-based solar power plants: both ideas have been around a long time, but both remain impractical for a whole variety of reasons related to cost, scale, and significant technical hurdles.

If the downsides of desalination—brine disposal, cost, and energy intensity—can be managed, then it is a promising source of freshwater for society, which can solve many water problems. However, as noted earlier, desalination can exacerbate energy problems, which invites these technical solutions (renewables, better design, etc.) to minimize impact.

If the challenges of desalination remain too vexing, then some communities can instead turn to harvesting water out of the sky. In particular, harvesting fog could be a ripe opportunity in certain parts of the world. There are some climates like northern Chile and San Francisco where freshwater is limited, but fog is abundant. In these locations, fog harvesters, which are mesh nets of tightly woven advanced fibers, can be erected vertically.[16] As the fog passes through these nets, water droplets collect on the mesh and run to the bottom where they can be collected. These harvesters might be able to collect twelve liters (about three gallons) per square meter of net per day, which is enough to meet the needs of an individual. That means that in a place like Chile, millions of people could have their water needs just by collecting a few percent of the fog's moisture. While those conditions do not necessarily apply in other parts of the world—and the chances of success from widespread implementation in Chile are still not known as it is in just proof-of-concept mode—every little bit helps.

In addition to switching the energy or water sources, it is also possible to implement advanced technologies that have enhanced performance. These include water-lean energy technologies, energy-lean water technologies, smart technologies, and distributed systems. Advanced technologies often have the drawback of requiring additional investment up front, in exchange for cost and resource savings downstream.

Switching to water-lean energy systems is particularly promising. Power plants can switch their wet-cooling systems to air-cooling or hybrid wet-dry cooling modes of operation. While that expense as a retrofit for existing power plants might not be cost-effective, for new power plants that might have trouble gaining access to cooling water, it can be a sensible approach to take. Because the heat capacity of air is one-fourth that of water, air-cooled systems must move four times as much air. That means the cooling systems need bigger fans and are much larger overall. Unfortunately, capital expenses scale with size, so the larger systems mean greater cost up front. At the same time, dry-cooling systems offer a performance disadvantage: power plants with dry cooling usually operate with an efficiency that is 2–10 percent lower than a similar system that uses wet cooling, depending on prevailing meteorological conditions.

In addition to dry cooling, there are also hybrid wet-dry systems. For example, some new designs from Johnson Controls use dry cooling in the winter, but wet cooling in the summer. In the winter, the air is cooler and usually the grid has sufficient power capacity available to meet demand. Because of the cooler air, the performance drawback from air cooling is not quite as bad as it would have been otherwise. And many power plants are looking to reduce their plumes of condensed vapor to reduce the visual pollution they cause. Switching to air cooling eliminates the plume and reduces annual water use.

In the summer, the performance cost of dry cooling is typically larger. At the same time, the summer is when the grid has less margin for error and so any performance cuts from power plants are felt more acutely. The hybrid system could be switched from air cooling, which works fine during the winter, to wet cooling, which boosts badly needed power output during the summer. In addition, plumes are not a problem in the summer as the hotter air temperatures prevent condensation of the evaporated water. In most places water availability is lower in the summer than the winter, so this arrangement would rely on saving up water during the winter months when air cooling is used so that the water is available for plant cooling in the summer.

The key tradeoff with dry-cooling systems is that they cause a persistent performance drag during normal operation, in exchange for performance resiliency during times of extreme drought or heat wave. Heat waves or droughts can force power plants with wet-cooling systems to draw down on their power output or shut off completely because of the risk of exceeding thermal pollution limits in the waterways or because of falling reservoir levels. Dry systems are less vulnerable to such problems. While the performance during a heat wave might be worse at a power plant, those with dry cooling are not subject to the thermal pollution limits because they do not dump heat into waterways. Because they don't require water, the drought does not affect their performance. That means, while other power plants in the grid are turning down or off, the plants with dry cooling would be able to continue generating power. At the same time, power prices are usually higher when the grid is strained. The dry-cooled power plants that keep operating while the others dial back will make a tidy profit.

In a nutshell, that means power plants with dry cooling would trade a persistent small percentage power loss in exchange for having steady performance at the same time their competitors might not be able to operate at all. If droughts and heat waves become more prevalent, then dry cooling would be a more attractive option economically. It is somewhat like fire insurance for homes. Fires rarely burn down our homes, but when they do, they cause a lot of damage. The same idea could be used to describe a power plant's vulnerability to being turned off from drought: it rarely happens, but when it does, the lost revenues are substantial. For the case with our homes, we pay approximately 1 percent of our home's value each year to buy fire insurance just in case that rare fire happens. We have determined that it's a lot easier to pay a predictable premium of $1,000–2,000 each year for insurance than to have the risk of paying $200,000–400,000 every fifty to one hundred years when a tragic fire happens. That math transfers over to power plants and their cooling: is it better to pay a 2 percent penalty in performance every year to avoid a 100 percent penalty that might happen only once every forty to sixty years? Our research at UT Austin suggests that it is worthwhile.[17]

However, switching the cooling system is expensive. In one case, state regulators required cooling towers to be installed at Oyster Creek, a 645-megawatt nuclear power plant in New Jersey, to mitigate environmental impact on Barnegat Bay. Estimates for the cost to convert the cooling systems ranged from $79 million to $801 million, which was prohibitively high. It ended up being cheaper for the operators to simply shut down the power plant rather than pay hundreds of millions of dollars for the cooling towers.[18]

While switching the cooling technology is an obvious way to reduce water use, switching the power cycle for the power plant also makes a difference. Coal plants use steam cycle designs with an overall efficiency of approximately 30–35 percent that do not compete well against more advanced combined cycle plants that have 40–60 percent efficiency, especially when the latter designs are coupled with cheap natural gas. Since the revolution of hydraulic fracturing, which accelerated starting in 2008, natural gas production in the United States grew quickly, pushing natural gas prices down from thirteen dollars per million Btu in the first decade of the twenty-first century to under three dollars per million Btu in the second decade. The century-long dominance of coal for electricity generation was threatened for the first time. For decades, coal has provided more than half of U.S. power generation, whereas natural gas provided about one fifth. However, in April 2012, when natural gas prices were about two dollars per million Btu, and overall electricity demand was relatively low, more electricity was generated by natural gas that month than from coal. That was the first time in U.S. history for such a phenomenon. Ever since the first coal plants came online, surpassing the hydropower capacity that had been built in the 1880s, coal has been the dominant power producer. While that month might have felt like a slight anomaly, the trend of decreasing power generation from coal and increasing output from natural gas is well under way and accelerated in 2015.

The increasing market share of those natural gas combined cycle plants comes primarily at the expense of traditional coal plants. This trend has several co-benefits: the newer natural gas plants are more efficient, produce fewer emissions, and require less water than the older coal plants. That means we can use plants that are clean and lean to replace those that are dirty and thirsty. While the water savings at the power plant are appealing, won't the water used to produce the natural gas in the first place undo all those benefits? We asked the same question in my research group, and a gifted student named Emily Grubert answered it.[19] She found that without a doubt, hydraulic fracturing for natural gas production requires more water per unit of energy than coal mining. However, she also found that the avoided water use at the power plant was much bigger than any increases at the point of fuel production. Those savings occur for three reasons. First, because the combined cycle plants are more efficient, they need less water for every unit of electricity produced. Second, the combined cycle is partially air-cooled, so even if the system had the same efficiency as coal, it would still save water. Third, because natural gas is much cleaner than coal, the water-intensive scrubbing systems to remove pollutants from flue gases could be avoided. Overall, the switch from a traditional coal plant to a new natural gas combined cycle can cut the life-cycle water intensity in half, in spite of the water requirements for fracking. This result was surprising to many, and I still get angry comments from people in the coal industry about that work.

Although this is good news about the life-cycle water intensity of natural gas—despite the water needs of the fracking itself—that doesn't mean the water needs of fuel production should be ignored. The extractive industries can try new techniques that do not require freshwater, such as waterless fracking or use of effluent or saline water instead. For waterless fracking, there are a few different approaches. Some technologies use nitrogen-based solutions instead of water-based solutions, and other approaches use propane gels to fracture the formations.[20] After injection into the shale, these gels evaporate into propane gas and escape out the well with the other gases that are being produced. The gaseous propane, which is a fuel, can be captured along with the other gases and used again in gel form or sold as a fuel. Although this idea has been demonstrated, some important risks of injecting flammable gases at high pressures deep underground still need to be managed.

Just as there are water-lean energy options, energy-lean water options can also be developed. Much of the energy intensity for water shows up in a few key places: water pumping, water treatment, and water heating. Water pumping is a pretty standard technology that has been around in a variety of forms for thousands of years. While pumps have improved significantly there is still more to do.[21] Large water pumps require a lot of power. It is not unusual for a large pump to require nearly two megawatts of electrical power, which is the same amount of power that a massive wind turbine can generate. A pump that size could move water at a rate of more than fifty thousand gallons per minute. The efficiency of these pumps can be improved greatly.

Most pumps run at a constant speed. At the same time, approximately 75 percent of pumps are oversized, which means they are bigger than they need to be for the task at hand and are consuming much more energy than is needed.[22] Incorporating variable speed drives (VSD) and variable frequency drives (VFD) allows for tighter control over pump operation, including the ability to tweak the flow rates to match the desired conditions. Most pumps are either off or on at full power. By using VSD or VFD to dial back on pump output to match what is actually needed, rather than just running at maximum power continually, much higher efficiency can be achieved: some reports indicate energy savings of 30–50 percent. This potential is particularly important for pumps that have variable-duty assignments—that is, they aren't needed all the time. Ramping them up and down as they are needed saves a lot of energy compared with letting them run all the time. Changing their operation will also extend their lifetime and improve reliability. Switching to pumps with this design also saves a lot of money on energy requirements for moving the fluids around: it is reasonable to expect a payback within two years for this technology. These variable pumps can also be used to improve the power quality of the electrical grid, which can become distorted as we increase the number of electric vehicles and solar panels in our neighborhoods.[23]

Other opportunities are to improve the membranes that are used for treatment, especially for desalination. The membranes for desalination work by allowing water to pass through while leaving the salts behind.

However, to use these membranes, the pressures of the water must be elevated above the osmotic pressures that occur from the salinity gradient. Better membranes could function at lower pressures, which would reduce the power requirements and therefore the energy consumption for the water pumps. Membranes also degrade from fouling. By designing the membranes with more advanced materials and more sophisticated structures, they will be better able to resist fouling, which means they will last longer and require less energy for manufacturing replacement parts and maintenance.

We can also use new approaches for water heating. Since nearly 4 percent of our annual energy consumption is just for heating water in our homes and businesses, improved water heating technologies represent a substantial opportunity to save energy.[24] Much of this heating is done with electric water heaters. While electric water heaters are very efficient at the point of use, sometimes exceeding 90 percent efficiency, the power plants that operate them have 30–40 percent efficiency. Multiplying the efficiency of the power plant and the water heater together makes the lifecycle efficiency for conventional electric water heating about 27–36 percent overall, which is not that great. By contrast, natural gas water heaters with 60 percent efficiency at the point of use might seem worse compared to the 90 percent efficiency of electric heaters, but because they avoid the inefficient power plant, they are much better in end-to-end efficiency.

It is unfortunate that the EPA's EnergyGuide labeling systems, which only identify end-use energy requirements instead of life-cycle energy requirements, have been nudging people away from gas water heaters toward electric heaters. If everyone in the United States who is currently using an electric water heater switched to a natural gas one, then we could save a nontrivial amount of energy per year. At the same time, for locations where the electricity is provided by clean sources—such as hydroelectric in Norway or the Pacific Northwest, or nuclear in France—switching from low-carbon electricity to relatively higher carbon natural gas water heaters would be a step in the wrong direction. When it comes to selecting the right water heating approach, technology and location matter.

We could also install point-of-use insta-hot water heaters, which heat the water on demand, usually with natural gas. This approach has several advantages. It is more efficient than electric water heating, and it heats water only when you need it, rather than traditional systems, which heat water around the clock even though it is needed just a few minutes or hours per day. Also, the homeowner never "runs out" of hot water, as it is continually generated, which is convenient when hosting guests and several showers and loads of laundry and dishes need to be supplied over a short span of time. Another advantage of the tankless water heaters is that they can be located next to the shower or sink. In many homes, the hot water heater is located on the opposite side of the house from the shower, which creates the undesirable situation that people let the water run in the bath for several minutes while waiting for the hot water to arrive. With hot water on demand right in the bathroom, this wasteful waiting game can be avoided. At the same time, tankless systems are smaller and more compact than large tanks, which saves space in the home. The downsides are their higher upfront cost—a standard tank water heater can cost $500 whereas a tankless one might cost $1,500—and the possibility that never running out of water will accidentally encourage people to take much longer showers or wash more laundry in hot water rather than cold water.

An even better option is to use low-grade temperature sources such as solar energy or waste heat in the house for water heating. Solar thermal systems for rooftop water heating are cost-effective and work well. Waste heat in the home can also be used. Our ovens, toasters, dryers, air conditioners, and refrigerators all generate waste heat. Waste heat recovery devices, which are small thermoelectric gadgets that convert temperature differences into electricity, are plagued by low efficiency, but as their costs come down and performance improves, they will become attractive options for houses. And, if we plumbed up our homes in a more efficient way, locating the sources of waste heat near where hot water is needed, then that waste heat could be used directly for water heating. The hot water pipe coming to our shower could pass by the back of the refrigerators, oven, air conditioner, or furnace before a tankless water heater in the bathroom does a final little boost of heating to bring it to the desired temperature. That approach would save energy in two ways. First, we would need less energy for water heating. Second, we would spend less energy on air-conditioning in hot climates to overcome the waste heat.

Another technical approach is to use distributed energy and water technologies, which might save water and energy, respectively. Large centralized power plants use a lot of water all in one spot and the large centralized water treatment plants use a lot of energy, also in one spot. By making them smaller, modular, and distributed around geographically, there is a chance that less water or energy will be needed. Or, even if they require a lot of water and energy, those requirements would be spread out rather than concentrated in a single location.

Distributed energy systems include rooftop solar photovoltaic panels that generate power for a home or building on-site. These panels do not require cooling water to operate. Other distributed energy systems might include microturbines operating on natural gas or propane, which are air-cooled. Fuel cell systems, which also do not require much water, are another possible solution.

Though the equipment costs of the distributed systems are generally two to three times higher per unit of electricity than conventional power plants, they offer some performance or environmental advantages. Solar photovoltaic panels are emissions-free, which is important for protecting air quality. And their production aligns reasonably well—though not perfectly—with peak demand in hot, sunny climates. Moreover, with sufficient energy storage for providing backup, solar panels can be used to keep the lights on and the air conditioner humming even during a power outage, on cloudy days, or at night. Batteries and flywheels can be used to store electricity, and chilled or hot water tanks can be used for storing thermal energy.

Natural gas fuel cells and microturbines can generate heat and power. The fuel cells are nominally cleaner than power plants as their low temperature operation avoids the formation of pollutants such as nitrogen oxides. Microturbines are also relatively clean, but not necessarily any cleaner than a natural gas combined cycle power plant. Their main performance advantage is that the gas grid, made up of gas pipes buried underground, tends to be less vulnerable to windstorms, trees, and critters than overhead power lines. Consequently, hospitals, city halls and other critical facilities look to these distributed systems as a backup to grid-tied power.

Because of the security and environmental advantages of the distributed energy systems, they are growing in popularity. They also bring along their low water requirements as a nice environmental co-benefit, whether it was desired or not.

Just as distributed energy systems can help avoid water needs, distributed water systems also can help avoid energy needs. In the current incarnation of the U.S. water system, most drinking and municipal water is treated at centralized large-scale locations, then pumped throughout the service area, including to far-flung suburbs.

On average the American consumer uses approximately 150 gallons per day of drinking water in their households. That water is typically pumped and treated centrally, then pumped again to distribute it to the homes. However, only a small fraction of that drinking water is actually used for drinking. The rest is used for washing, watering, and cooking. So another option is that *only* the drinking water is centrally treated, and the other water is harvested and prepared in a distributed rather than centralized way. Instead of water treatment facilities needing to procure and treat 150 gallons per person per day, they could do that just for 25–50 gallons per person per day. That means less energy will be required as there will be less water to treat and pump over long distances. The key here is that the water we need for showering, toilet flushing, and clothes washing does not need to meet the same standard as for drinking.

The other water that we use could be provided by distributed water harvesting and collection systems at the household or neighborhood scale. That water needs less treatment, and since it would be closer to the end-use it could avoid a lot of pumping costs. This solution makes sense only where there is local water like precipitation to harvest, so many desert communities might not be able to pursue this approach. And, just as for distributed energy systems, the distributed water systems might be more resilient compared with the concentrated systems that could be taken off-line by a targeted attack or unfortunate natural incident.

Centralized water treatment has distinct energy-efficiency benefits from economies of scale that can be achieved with larger systems. Therefore, distributed treatment of drinking water actually might increase the energy consumption per unit of water. However, those efficiency losses from smaller treatment volumes might be overcome by the energy savings from shorter pumping distances.

Putting these together, it is possible to imagine new concepts for the built environment in which homes harvest as much water and energy on-site as possible, relying on the centralized grid as little as possible. That means capturing and reusing graywater and rainwater on-site while also producing and reusing thermal energy on-site, both of which offer savings. On-site electricity generation avoids the water embedded in grid-tied power plants, and on-site water harvesting avoids some energy embedded in pipe-tied water treatment plants.

Another form of distributed water systems is on-site water treatment for the energy sector. The large volumes of water produced and used by the extractive industries represent an interesting opportunity to close the loop. Coal production generates water that gets pumped out of the mines, and hydraulic fracturing can generate millions of gallons of wastewater at each drilling pad.

Because the wastewater volumes are so large, they can end up being a constraint on production. And, trucking or piping such large volumes off to an industrial wastewater treatment plant can be prohibitively expensive. However, distributed on-site treatment using new processing technologies can save water and energy. If properly implemented, these mobile treatment systems reduce the contaminated wastewater stream volumes by an order of magnitude. Subsequently, very little water needs to be trucked away for additional treatment, and the cleaner water that is left behind can be used again. This distributed system avoids the energy of transporting so much water and enables additional energy production. In this case, the distributed system wins on multiple counts.

Among the many technologies, the suite of so-called smart technologies offers some hope for mitigating some of the worst problems in the energy and water sectors. Today's dumb systems can waste water and energy without the wasters realizing it. Making matters worse, dumb systems are not resilient and are ultimately expensive. The idea of moving toward smart technologies, including ubiquitous sensing and smart meters for electricity, gas, and water to give more finely resolved information is appealing.

The most common type of electricity meter is a device that measures energy in kilowatt-hours through electromechanical induction. These meters were first demonstrated in the late 1800s. Remarkably, today's meters operate on the same principle. The electrical current spins a metal disc at a rate proportional to the electrical power—this is the familiar rotating dial inside the glass bowl of conventional electricity meters. It rotates slowly except for when large appliances are operating, and the meter counts the rotations to track total usage.

These meters must be recorded manually by meter readers, which is inconvenient since the meters are often located behind a fence at the back of the house, above the thorn bush and protected by guard dogs. That our meter technology is so antiquated and requires an inefficient, labor-intensive process for meter reading is startling. Making matters worse, electricity meters slow down with time with a phenomenon known as "creep." Cash-strapped consumers might appreciate these slow meters, because it means their bills are lower than they would have been otherwise. But it also means they—and the utilities sending them bills—are getting inaccurate information. Consequently, when new "smart" meters are installed, which probably have been recently calibrated, customers' bills can go up because they are accurate for the first time in decades.

Water meters use a simple rotating vane technology that records cumulative usage and is also at least a century old. Similar to the electricity meters, they need to be read manually and slow down with time. Though their location, typically near the curb, is more convenient than the electricity meters, the process still requires the expense of rolling trucks to send out the meter readers. Natural gas meters are similar to water meters, but because gas is so combustible, gas meters in seismic zones have automatic shut-off valves to prevent explosive consequences after earthquakes.

The water, gas, and electricity meters share some of the same problems. One of the biggest shortcomings is that their measurements and billing are disaggregated from use and time. Monthly bills include consumption information that spans a thirty-day period. Processing, printing, and mailing those bills takes another one to two weeks, which means that by the time consumers actually read their bills, the consumption information they receive includes activity from as much as five to six weeks earlier. It's pretty straightforward to assume that busy consumers might have trouble remembering what they did with their appliances forty days in the past. Water bills issued quarterly (not an unusual occurrence) exacerbate the problem even further. The complexity of the bills, cluttered as they are with separate taxes, fees, riders, and other costs not directly related to consumption, impairs the ability to interpret total usage. Few people have any idea how many kilowatt-hours of electricity, cubic feet of gas, or thousands of gallons of water they consume in a month.

Because the bill includes consumption information for the entire billing cycle, consumers are operating blind. The bills typically do not differentiate by time of day, date, season, household occupant, or appliance. That would be like shopping for groceries, but with no prices on any of the food items. With no price signal to steer our behavior, we would load up our carts with whatever looks appealing—steaks, specialty chocolates, and other high-priced items—leave the store, and take the groceries home, repeating that cycle twice weekly. We might even buy more than we need, throwing away the rest since for all we know it's free. At the end of the month, the grocery store would send us a bill for all the food we had purchased. Imagine our sticker shock when we see the tally. If we had price information about each individual food item, then we could shop in a more intelligent way, buying only what we need, wasting less, and prioritizing the more affordable items.

This preposterous grocery story scenario is similar to our dumb electric bills. Without price signals on individual appliances, it is easy to waste electricity and to accrue expenses without realizing it. Consumers might understand the bill's basic conclusions about their overall usage patterns, but might not know which particular appliances drive the consumption. If they knew how much energy their major appliances require, consumers might turn them off when unused or seek alternatives such as hanging clothes in the sun to dry instead of using the clothes dryer.

If we think the power sector is dumb, the water sector is even dumber. Water meters are often read quarterly, and in some municipalities such as Sacramento and Las Vegas there were no meters at all until recently, despite being in the middle of a desert. Wastewater is not measured, either. Rather, some utilities measure water use in the winter, assume all of this water is used for indoor purposes, as opposed to irrigation, and that all of that water makes its way down the drain, and then use that as the baseline estimate for the wastewater bill. That baseline is used for the rest of the year, whether irrigation is performed or not.

The idea that prices are the same for every appliance, every hour of every day of the year is astounding, as the rest of the supply and demand fundamentals of the market change every few seconds throughout the year. In the United States, electricity demand is higher in afternoons than nighttime and higher in summer than the winter. Water use is higher in the summer for irrigation than winter and higher during the day than night. Wind and water are both more abundant at times other than when demand for them is highest. Natural gas prices also have seasonal cycles based on demand for winter heating and summer power plant operation. However, for over a century our billing systems have not taken that information into account. Making matters even worse, some homes do not even have water meters installed; their bills include a flat fee based on the cost of infrastructure. These flat-fee bills actually encourage more consumption, since there is no price penalty for profligate or wasteful use of water.

This lack of information is particularly surprising given the revolution in the 1990s that dramatically reduced the costs of information technologies and opened up the pathway for ubiquitous sensing. It seems like the information revolution overtook every industry except for energy and water utilities. However, with the prospects for embedded information, the possibility to collect significantly more data about resource usage is tantalizingly close.

For electricity, gas, and water meters, it would be valuable to gain information about usage by time of use and the consumption of particular appliances. For water it would also be valuable to know the different types of water that are used: indoor versus outdoor, heated versus unheated, treated water versus graywater and blackwater, and piped water versus water collected on-site.

All this new information would be very useful for utilities and consumers. Consumers could track their own consumption and could also potentially track particular household members, which might be valuable for parents trying to convince their teenagers to take shorter showers. One friend of mine, Brewster McCracken, who is the CEO of Pecan Street, a nonprofit consortium conducting large-scale smart-grid and smart-water experiments, had a clever idea. Because his son has a propensity to take very long, hot showers, Brewster thought it would be a great idea to hang a timer in the shower that relayed price data from a smart meter. After some specified time limit, the readout would start counting down his son's allowance, which would be taxed to pay for the extra hot water.

In addition to helping parents rear resource-conscious children, that information could be used to spot leaks or broken appliances. Utilities could use that information to respond more quickly to outages, to predict maintenance issues, and to cut costs by automating meter reading through telemetry rather than requiring employees to read every single meter manually. Rather than sending out hundreds of meter readers, a van could slowly drive through the neighborhood picking up meter signals along the way. Homeowners would benefit by fixing problems before they get worse, reducing the risk of fire in case of gas and electricity problems and before serious household damage occurs from water leaks.

The real-time information could be used to balance supply and demand, which is particularly relevant for the electricity markets. For on-site generators such as rooftop solar panels, the smarter meters could track how much power homes provide back to the grid to account for credits that would be due.

Smart meters and appliances will bring forth smart billing and smart pricing, too. In the United States, prices for electricity should be higher during summer afternoons, and lower in the middle of the night. In France, electricity prices should be higher in the evenings in winter when electrically heated homes cause demand to spike. Water prices should be higher in the dry seasons than wet seasons. With time-resolved metering, utilities could also implement time-of-use (TOU) pricing. These price tiers could have off- and on-peak price signals to shift demand across the hours and days. That also means regulators who use century-old business models to approve fixed prices will also have to become smart to allow utilities to shift to these newer pricing schemes. It is worth noting that most European countries, which many Americans decry as socialist, already have widespread time-of-day pricing so that they can exploit the power of markets to achieve efficiency. By contrast, capitalist America typically has fixed-price, infrastructure-based pricing. Leave it to those clever socialist Europeans to use market forces more effectively than capitalistic Americans.

The implementation of smart meters enables the installation of smart appliances, which might be used by utilities for grid balancing. Nonessential appliances such as pool pumps or hot water heaters could potentially be turned off automatically during peak loads or high price times of day to save consumers money. There are even some designs for smart clothes dryers that keep the tumbler rotating but turn off the heating element periodically when the grid needs to reduce demand. In France, there are more than 13 million smart electric water heater units installed. They have a peak demand at night in the winter of eight gigawatts, which is equivalent to the output from seven nuclear power plants. By remotely turning them down when they are not needed, three gigawatts of consumption can be avoided.[25]

These new technologies also democratize the infrastructures, allowing third-party vendors into the system who will sell services that help consumers reduce their expenditures. These energy service companies (ESCOs) help instrument and manage a building especially for the commercial sector, reducing consumption and saving customers money. The ESCOs earn revenue based on a portion of the savings—the more they save their customers, the more money the ESCOs earn. But such business models are hard to implement with the existing dumb systems. For the residential sector, opening up the data streams means consumers might be able to buy a clear, digital readout to install in their kitchen with real-time, appliance-specific, and cumulative consumption information—much better than the slowly spinning analog dial that is outside at the back of the house and hard to read anyway.

The smart systems still have downsides, and we would be wise to prepare for them. There are the security challenges: to security experts smart devices look like entry points into our critical infrastructures, making them vulnerable to cyberattacks. This issue is real and worthy of further consideration. However, one counterargument is that a web of smart devices coupled with distributed energy and water systems would be inherently more resilient and able to respond more quickly in the event an attack actually occurs. Rather than providing a target of just a thousand power plants and thousand water plants for the existing, centralized system, the smart, distributed system would have millions of different devices, making it harder to take them all off-line in a single fell swoop.

Another disadvantage is the looming burden of "big data." According to humorous observers, big data is like teen sex: everyone is talking about it, everyone thinks everyone else is doing it, but in truth we are all just clumsily fumbling around in the dark trying to figure out how to make the different pieces go together. Today's dumb system has the advantage of including very little data, which means there is not much information to collect, store, archive, or retrieve. Conventional meters have two data points for a billing cycle: one data point at the beginning of the month, and another data point one month later. The total consumption is simply the difference of those two points. A smart system will have more frequent readings: with a fifteen-minute interval between readings—which is typical—meters will produce about three thousand meter readings per month. But, utilities, working with decades-old software and computing systems, are often ill-prepared for such a transition to the twenty-first century. They often do not have the skilled personnel, the data, the computers, or the cultural mindset to use these datasets, much less make them available to consumers.

For one case study, Austin Energy, a municipal utility on whose governance board I served for five years, completed a rollout of smart meters to more than 300,000 customers from 2008 to 2010. Austin Energy is often considered one of the nation's leading and most forward-looking utilities, and so it is not surprising that they were one of the first major municipalities to complete the installation of so many smart meters. The whole effort cost a little over $150 per meter, or about $50 million in total. After quickly installing the meters they discovered that their antiquated data systems, not to mention their personnel in their customer service billing divisions—simply were unprepared for such an event. The billing systems were not upgraded until more than two years later, in late 2012, at a surprising cost of an additional, unexpected $50 million. Handling the data cost just as much money as installing the systems themselves. Big data is liberating and enabling for the consumers and utilities, but is also expensive and cumbersome to manage.

In some ways, the smart revolution is an allegory for everything else about these infrastructures: it is a solution to one set of problems that introduces a new set of problems. And so the cycle will begin again.

Some of the smart grid benefits for electricity might also be relevant for the world of smart water. My good friend Amy Hardberger is a water lawyer and a collaborator for some of the research discussed in this book. She is an active environmentalist who practices what she preaches, attentive to her resource consumption and aware of the impacts to the environment.

One day in the late 2000s, while working in her yard, she discovered her water meter buried a few feet underground. She naturally wondered how her water bills were generated if the meter was located several feet underground. She called to inquire about how the water utility determined the amount of her bill each month.

The water utility replied that a meter reader goes by each house to determine the amount of water that had been consumed, which sets the bill. While that answer would have been true in many corners of the United States, Hardberger laughed, knowing that the answer was wrong, as her meter had been securely buried underground. Presumably the meter reader had been making up numbers for years to make up for his or her inability to actually read the meter as that would have required a serious digging exercise. Knowing that a high bill would raise concerns, the reader probably cleverly skewed the numbers low, figuring that a low bill would not inspire complaints by the homeowner. That Hardberger was a prudent water user was coincidental: for her, low bills would have been appropriate, as she consciously kept her water use to a minimum.

This anecdote reveals just how dumb our water infrastructure can be. But the power sector is also considered dumb, as very little information is embedded with the end-use consumption. Many consumers such as tenants in apartment buildings have absolutely no idea how much electricity they consume, if the utility charges are included in their rent. In many cases, individual apartments or condos do not have independent meters, as all the consumption is aggregated into a single meter for the entire building. Further, in many municipalities, that bill often aggregates several services, including water, wastewater, and trash disposal. That means customers don't have good information about their consumption, but it also means that many leaks and losses go undetected.

I have my own experience with dumb meters, which is slightly embarrassing because I try to be attentive to energy and water matters. My house has very hard water that causes buildup in the pipes and tends to be damaging over time. In mid-October 2010, that buildup caused a break in a hot water pipe in our home's concrete foundation, but because of a dumb water meter, a dumb electricity meter, and a dumb homeowner—me—I did not figure out that we had a leak for two and a half months.

It started when my wife and I noticed that there was a spot in our concrete floor that was a little warmer than normal. We didn't really know what to make of it. Our electricity and water bills came a few weeks later and I didn't notice any variability beyond the normal range. To be honest, I was not looking at the bills with any more scrutiny than normal, but even if I had, since bills change month to month anyway, it's not clear what I would have seen.

As the weeks rolled on, the warm spot in the floor didn't go away. Our water usage was about 50 percent higher than normal, but since most of a water bill is composed of fixed monthly costs and very little of the bill depends on water consumption, the total amount of the bill didn't change much. Our electricity bill didn't change, either. While we have a solar thermal system for our water heating, because the water was being used continually, the electric backup for the heater would have been triggered. We were undoubtedly spending more electricity heating water, but since it was in the fall, our use of air-conditioning was dropping. Despite the hot water leak and the additional electricity consumption for water heating, our electricity bill actually dropped overall.

Finally, yet another month later, the spot in the floor was getting progressively warmer, a crack emerged in the floor by the back of the house due to shifting soils from extra water, and our third bill arrived—a full two and half months after we first noticed the warm spot. By this point we could hear water rushing. I hate to admit how slow we were to figure out what was happening, but it finally dawned on us from all those separate points of evidence that there was a potentially expensive and damaging hot water leak *inside* our concrete slab.

We called the plumber and discovered we were lucky. First, the slab wasn't damaged, which is remarkable. Repairing a damaged slab can easily cost tens of thousands of dollars. Second, the pipe could be fixed through a bypass in the wall, which only required the removal of some sheetrock rather than drilling or cutting into the slab, saving us many thousands of dollars.

While smarter homeowners could have avoided this risk—the irony of an energy-water nexus researcher getting hit by an energy-water nexus failure without noticing is worthy of some sort of parody—a smarter system could have spared that wasted water and electricity and avoided the risk of expensive damage and repairs. A smart water system could have alerted us that flows out of the hot water heater were much higher than normal and were occurring around the clock, rather than just when hot water appliances and fixtures were being used. A smart electricity system could have alerted us that the consumption for the hot water heater was much higher and steadier than normal. Smart meters would have given us data every fifteen minutes instead of every month. Thus, instead of needing two and a half months to get three data points about our use, we could have been alerted within a few hours.

Such a smart system would have been very advantageous. But, like most Americans, we did not have such a system. Scaling that idea larger, the utilities would benefit from having sensors and meters that automatically alert them to power outages and leaks. Today they rely on people to call the utility when there is an outage. In those cases, based on the location of who calls in, they can identify the affected area and search for the problem by sending crews out with trucks and flashlights. With a smart system, the utility would know the exact location of the break, reducing outage times dramatically.

While the fancy bells and whistles of smart water and advanced technologies are appealing, some low-tech solutions can also be very effective. One of those ideas is the WaterWheel, which is a round container that holds fifty liters of water and can be rolled like a wheel.[26] For poor people who do not have the energy for piped water and have to collect water manually—namely, women—the wheeled container means they can roll the water from the well to their home rather than carrying it on their heads, reducing the burden. While this is not quite as convenient as a modern piped system, it is a step in the right direction.

We can also try cross-sectoral problem solving. Although the water sector's problems can become the energy sector's problems and vice versa, the corollary is also true: each sector can integrate solutions that mitigate problems in the other. We can use the water sector to produce energy and the energy sector to produce water.

One useful example is the approach of turning wastewater treatment facilities into producers of energy rather than consumers of energy. A nontrivial amount of energy is required for wastewater treatment facilities. That energy goes for a variety of purposes such as pumping, filtering, stirring, and ultraviolet irradiation. Consequently, energy expenses are the primary operational costs for the treatment plants. As treatment standards get stricter, the energy requirements typically increase.

However, wastewater treatment facilities also circulate large volumes of water that have a significant fraction of organic content. If the organic-rich wastewater sludge is sent through anaerobic digesters, then the decomposition of the sludge would produce biogas, a mixture of methane and carbon dioxide. This process also produces a solid digestate that can be used as a fertilizer. Anaerobic digestion is not that controversial as a treatment approach, as it is already implemented around the world. But unfor-

Wello's WaterWheel is one low-tech innovation—the water vessel is also a wheel—that will reduce the burden of manually fetching water. [Photo courtesy of Wello]

tunately, much of the biogas that is generated is either flared or vented: few facilities actually put the biogas to work.

Facilities that capture and use the biogas for process heating or on-site electricity generation will reduce their energy bills, though there are additional capital costs for the digester equipment if the utility did not already own them. It is even possible for many plants, especially those with larger flows that have more potential for generating biogas, to become nearly energy self-sufficient.[27]

In Austin, the use of digesters produces solid waste materials that are then mixed with plant trimmings to form a valuable soil amendment known as Dillo Dirt. "Dillo" is short for "armadillo," which is a prehistoric-looking creature native to central Texas. The city sells the soil amendment to landscapers, farmers, and home gardeners. Doing so helps close the loop, turning the waste stream into a valuable commodity, while at the same time generating revenues for the wastewater treatment facility. The sludge-to-biogas-to-fertilizer approach serves two purposes: saving money and saving energy. Because fertilizers are energy intensive to manufacture—they are typically made by converting natural gas into ammonia—each pound of Dillo Dirt fertilizer produced from wastewater saves energy. That means wastewater treatment plants can go from consuming energy for their operations toward producing energy in the form of biogas and saving energy elsewhere.

While Dillo Dirt is a good news story, it is not without its controversy. Two of Austin's gems converged in a messy story in October 2009. *Austin City Limits* is a live music show that has run continuously for decades, making it one of public television's most successful programs of all time. It is also an Austin landmark of sorts, as it affirms Austin's self-declared designation as the live music capital of the world. In addition to the television show, filmed before a live studio audience, the Austin City Limits festival is attended by nearly a half million live-music lovers over several days. It takes place in the fall at Zilker Park, a large and iconic urban park along the south shore of the Colorado River that winds through downtown.

The fall weather in Austin is unpredictable, varying from hot to cold, wet to dry. After several years in a row of dry weather, Zilker Park's lush green fields turned to dust, causing massive dust clouds for the 2008 festival, which took away from the normally joyous atmosphere. To prevent that problem from happening again, the festival organizers and the city of Austin resurfaced Zilker Park's fields in advance of the festival in 2009. In doing so, they amended the soils with Dillo Dirt. As fate would have it, that year would be remarkably wet, with epic rainstorms occurring during the festival. The formerly dry, dusty fields turned into boggy fields of gooey mud several inches thick. People were losing entire shoes and other garments and personal items in the quicksand-like fields. News spread quickly that the soils and mud were laced with Dillo Dirt. Festive partygoers who reveled in the idea of frolicking in the mud in some sort of mass remembrance of Woodstock decades earlier quickly became disgusted when they realized that they were shin-deep in human waste. Treated human waste, to be sure, but human waste nevertheless. The initial outcry, even for Austin, was notable. No one died and Dillo Dirt was declared safe by public health authorities, but this episode reveals that even in environmentally conscious Austin, recycling our wastes for useful purposes can still hit some roadblocks.

Just as the water sector can be used to generate energy, if designed the right way, the energy sector can transform from being a water consumer to a water provider. Some of the water that oil and gas operations produce is mildly brackish and could be used for industrial applications such as power plant cooling. Or the produced water could be collected and used again at neighboring sites. If combined with distributed wastewater treatment technologies, the oil and gas sector could even generate freshwater on the drilling pad. Scaling up this idea, oil and gas companies might someday become oil, gas, and water companies. Instead of spending money to get water, they will earn money to provide water. The oilfield service companies—Halliburton, Schlumberger, and National Oilwell Varco—are already thinking about this, as water management is one of the biggest headaches at the well pad.

One opportunity would solve three problems at once. The shale revolution has created a remarkable upswing in domestic oil and gas production. It has also produced three significant environmental drawbacks: increases in demand for freshwater for hydraulic fracturing, produced water with high levels of solids and undesirable constituents, and flaring of natural gas. Gas flaring is particularly common in places like the Bakken Shale in North Dakota and the Eagle Ford Shale in south Texas, which produce a lot of gas along with the oil, but lack sufficient gas pipelines to move the gas to market. Gas flaring in Texas more than tripled between 2010 and 2014, and about a third of the natural gas pulled out of the ground in North Dakota is flared. Flared natural gas is an environmental challenge because it has all the pollution of combusting fuels without harnessing useful services. The water competition for fracking and produced water are also challenging for all the reasons discussed previously. But by putting together the two waste streams—flared gas and produced water—a new source of water can be supplied, reducing competition with other sources. Flaring is particularly prevalent during the first few weeks of a well's completion. That is also when the wastewater flow rates are highest and the levels of dissolved solids are lower. Instead of being flared, the gas can be used as energy to treat the water, turning the two waste streams into a useful commodity of water that is clean enough to be used again. In Texas, we found that if just the gas from the first ten days of flaring was used to treat the wastewater, it would increase statewide water supply by 1–2.5 percent.[28] Since many of the shale basins are in water-stressed areas, this solution would be particularly useful for mitigating water strain from increased shale production.

The power sector can also be used to generate water by coupling power plants with desalination systems. This idea was flagged earlier with the possible switching to saline or brackish sources, which can be performed by coupling the water treatment system to a wind or solar power plant. Doing so solves several problems at once. Wind and solar farms are plagued by intermittency, as they only work when the astronomical and meteorological conditions are favorable. This intermittency causes problems for grid management and is one of the most important downsides of those renewable sources.

This intermittency is both predictable—the sun not shining at night, for example—and unpredictable, such as that due to sudden weather changes. On the other hand, desalination of brackish and saline sources is very energy intensive and therefore carbon intensive because of its reliance on today's fuel mix. Integrating the two allows them to help mitigate each other's problems. The water treatment process itself can be operated in an intermittent way to match when the wind or sun are available. That is, instead of changing the power plant's supply up and down to match the demand from the user, the demand can be turned up and down to match the supply. As wind and solar vary hour by hour, the desalination plant can follow it, consuming the power when it is available and idling itself when power output is low. Because wind and solar are low-carbon sources, they help mitigate the carbon emissions from desalination that is powered from the conventional grid.

In what can only be described as a cruel joke by Mother Nature, some of the world's major solar-rich deserts overlay abundant brackish groundwater, as in northern Africa. There are also abundant solar and wind resources in desert areas of the United States, such as New Mexico and Texas, where significant resources of brackish water aquifers are located. In the southwestern United States, the co-location of wind, sun, and brackish water might make desalination a more attractive option economically and environmentally. And, there are locations where the brackish aquifers are fairly shallow, the water is not very salty, and wind or solar resources are abundant. In those locations, it is more profitable to use the renewable energy to generate freshwater than to make electricity. Such a coupled system takes two low-value products—brackish water and intermittent renewables—and turns them into a high-value product (treated drinking water).

Research performed by my students found that in areas with a lot of wind and solar resources, large volumes of shallow and mildly brackish groundwater, and high local water prices, an integrated system may significantly improve the supply of water and electricity and is potentially more profitable than a standalone power plant.[29] An integrated facility would require a large up-front capital investment. But, then it would be more flexible, as it could produce electricity and water, and it would be more efficient overall. Plant operators would benefit by hedging the power and water markets, selling whichever commodity is most profitable at that particular moment. An integrated plant would sell water when water prices are high and would sell electricity when electricity prices are high. It could easily store water when conditions are mediocre for its sale. Storing desalinated water could act as a proxy for energy storage. Notably, it is a lot cheaper and easier to store water than it is to store electricity. Instead of expensive lithium-ion batteries, for example, you can simply dig reservoirs or use holding tanks.

Because groundwater is typically cooler than surface conditions in warmer climates, it could be used to cool solar panels. While solar panels do not require cooling, doing so improves their efficiency. In the process of cooling the solar panels, the brackish water is heated, which improves its membrane throughput and recovery during the desalination process. It is a win-win scenario, with both water and electricity being produced with higher efficiency.[30]

In addition to wind and solar providing electricity, they can also provide direct inputs to drive the process. Windmills, instead of generating electricity, could provide direct mechanical power to operate the pumps that push the brackish water through membranes. And solar systems, instead of generating electricity, could generate the heat that is used to boil the brackish water. The energy requirements of the first approach depend on the salinity of the water, and the second approach depends on how hot the water feed is.

One of my students conducted a global examination of where solar-based desalination makes sense and could potentially be done sustainably.[31] She used several criteria: locations where freshwater is scarce, urban populations exceed one million people, it is sunny, and the water is warm but not too salty. The thinking is that people would not be willing to endure the expense of a desalination system unless water is scarce, and that only large municipal areas could accumulate the necessary capital to build the system. Once in place, those systems are more efficient if the water is warm, which helps it flow through the membranes; the water isn't too salty, which reduces the energy required for desalting; and sunshine is abundant. This research determined that there are indeed places throughout the tropics where water and radiation conditions are well matched for solar-powered desalination: the Bay of Bengal and other tropical locations with large cities near the coast met the criteria. The same kind of analysis could be done for wind-powered desalination, identifying water-stressed cities near the coast where wind is abundant and the water isn't too salty or cold.

Another approach is to use the waste heat from the conventional power sector to perform desalination. This is the approach used in Abu Dhabi, where desalination plants are coupled to power plants. The power plant runs on natural gas, and the waste heat boils an incoming stream of seawater. This integration makes perfect sense for a place like Abu

Dhabi, which is energy rich and water poor: trading energy for water is a very rational thing to do. And using the waste heat for the desalination is an efficient design from an engineering perspective as it saves significant energy for producing freshwater.

One surprising outcome from Abu Dhabi's arrangement is that it produces extra water in the summer and waste electricity in the winter.[32] In the summer, when desert temperatures are soaring, the electricity demand for air-conditioning is very high. Because the water production is tied to electricity production, as more power is generated, water is produced, too. High electricity demand in the summer means excess water production. Unexpectedly, there is more water available in the summer than the winter when power demand is relatively lower.

As there is no place to store it with the existing infrastructure, that water is used for seemingly profligate purposes such as irrigating golf courses and growing crops that are not native to the desert. Maintaining these golf courses and crops keeps the need for freshwater high in the winter. As a result, the power plants are turned on just to make the heat to desalinate water to meet this demand.

From an engineering perspective, the integration of power production and freshwater treatment is nominally a good idea that yields performance improvements. But, including the human element, this coupling causes wasted water in the summer (because of excess electricity consumption) and wasted electricity in the winter (because of excess water consumption).

That anecdote is a good reminder that taking a holistic view for resource planning is the critical missing piece. Systems-level thinking and integrated designs can improve efficiency, reliability, cost, and robustness. Unfortunately, political roadblocks and isolated decision-making often make those approaches difficult to implement. Those technical solutions still are not enough, as we hit human barriers with bad policies, poorly designed markets, and cultural indifference. We need nontechnical solutions, too.

NINE

Nontechnical Solutions

IN ADDITION TO THE TECHNICAL solutions for meeting our expanding needs for water and energy, there are a host of nontechnical solutions that include policy choices, economics, cultural forces, behavioral changes, and markets. These solutions include investing in research and development (R&D) for energy and water, setting strict performance standards, and designing more functional energy and water markets. Above all, conservation is the most promising path forward.

One important option is to implement policies that support massive investments in water and energy R&D on a scale that makes a real difference. In light of woefully low investments in R&D over the decades for both sectors, a commitment on the scale of a Manhattan Project or Apollo Project for clean energy and water might trigger a substantial breakthrough.

During World War II, the Manhattan Project was a national priority. Significant increases in scientific-industrial-military research led to the development of a nuclear weapon that could end the war. The Manhattan Project at its peak was responsible for about $10 billion per year of R&D funding in the United States, which was a sizable fraction of all R&D in the nation.[1] It was an overwhelming diversion and concentration of resources with a singular aim of achieving a breakthrough that would change the world. And it achieved its goals.

A similarly ambitious call to arms for concentrated R&D to achieve an ambitious purpose was with the space race of the 1960s, peaking with the Apollo Project to send men to the moon and back. From 1964 to 1967, space exploration received the lion's share of funding, with over 60 percent of the total nondefense federal R&D budget for the Apollo Project. The space race was a top national priority, and that investment not only put human beings on the moon but also stimulated progress in a number of technologies like miniature electronics and computing that are still producing economic benefits.

In May 1961, President John F. Kennedy challenged the nation to win the space race in a famous speech before the U.S. Congress: "No single space project in this period will be more impressive to mankind, or more important for the long-range exploration of space; and none will be so difficult or expensive to accomplish." In response, by 1969, we put humans on the moon and returned them safely to Earth. What most people don't realize is that in press conferences and campaign speeches, Kennedy made a similar challenge to the one that kicked off the space race, urging the United States to develop desalination technologies. "If we could ever competitively—at a cheap rate—get fresh water from salt water, that would be in the long-range interest of humanity and would really dwarf any other scientific accomplishment."[2] Kennedy understood that improving desalination technologies holds the prospect for raising people out of poverty and improving public health globally. Unfortunately, that challenge failed to spur the nation to action.

Despite Kennedy's call, water R&D investments are a small fraction of total U.S. R&D spending. They are so small, in fact, that they have not been tracked by the U.S. government, making it difficult to provide firm estimates of total spending for many years. There is no federal Department of Water, and that might be part of the problem. Even though there is a federal Department of Energy, energy R&D investments are also relatively small compared with health, space, and defense research. That phenomenon is ironic, since it can be argued that effective R&D and implementation of new energy systems could reduce our health problems, improve our national security system, and enable our space program. Energy R&D spending peaked in the 1970s in absolute and relative terms at nearly $8 billion of annual investment in response to the two different energy crises.

Spending grew toward the end of the Ford administration and continued through the Carter administration, until it shrank under President Ronald Reagan. The federal energy R&D budgets continued to decline under Presidents George H. W. Bush and Bill Clinton, dropping significantly during the late 1990s when oil prices were very low, energy crises seemed a distant memory, and balancing budgets was a policy priority. Funding held relatively steady (or even increased slightly) at approximately $1 billion annually during President George W. Bush's first term, though the priorities shifted from alternative fuels and efficiency technologies toward nuclear and fossil fuels production. Funding nearly doubled to $2 billion toward the end of his second term, partly in response to the oil price spikes in 2005 and 2008. Keep in mind that this total funding—$2 billion from federal sources, and if you include all other governmental spending on R&D still less than $4 billion—is less than what *one* pharmaceutical company would spend on its R&D division.

Under President Barack Obama, energy R&D more than doubled again, exceeding $4 billion annually, as part of the American Recovery and Reinvestment Act and other stimulus provisions that were executed in response to the economic collapse of 2008. Those investments prioritized alternative and domestic energy sources and included significant funding for large projects, often in the form of cash grants and loan guarantees. Whether energy R&D will continue on this path is hard to foresee, but it is a sign that the nation takes its energy problems more seriously now than in the previous decade.

If energy R&D is too small, water R&D is even smaller. Neither desalination nor the larger issue of the nation's water infrastructure has received much public attention or regular directed federal support. Water R&D has not been a consistent priority, and investment has endured erratic boom and bust cycles. Sheril Kirshenbaum, a highly regarded science writer, captured this phenomenon in a piece we wrote together titled "Another Giant Leap for Mankind."[3] Initially the nation responded to Kennedy's comments about converting saltwater to freshwater. During the 1960s and 1970s, the U.S. government cumulatively spent over $1 billion (without adjusting for inflation) on desalination R&D alone. The Water Resources Research Act of 1964 led to the creation of the Office of Water Research and Technology in the Department of Interior in 1974 to promote water resources management, and it also helped to establish water research institutes at universities and colleges. Three years later, the Water Research and Conservation Act authorized $40 million for demonstration-scale desalination plants. The following year, the Water Research and Development Act extended funding through 1980.

But the Office of Water Research and Technology did not last. Just eight years after it opened, the Reagan administration abolished it, distributing authority over water programs among a host of agencies and departments, making it extremely difficult to track government R&D spending on water. Because of this low level of funding, it is no surprise that water treatment technologies have evolved so slowly, that water infrastructure leaks so abundantly, and that water quality is at risk from a variety of societal activities and policy actions. Despite decades of building the world's greatest innovation and R&D system, U.S. progress in water innovations seems halting and stunted, especially when compared with the advances that occurred in parallel for information technology, energy, health care, or just about any other sector critical to society.

As Kirshenbaum likes to say, "Just imagine what we would have accomplished by now had we devoted the same attention to looking for water on Earth as looking for water on the moon." Even the youngest Americans can quote Neil Armstrong, but it is hard to imagine children today recounting the name or accomplishment of a water scientist's great laboratory triumph or the name of the British doctor who identified cholera as a water-quality problem. The article Kirshenbaum and I wrote said: "We celebrate the space program as one giant leap for mankind. Now it's time to take a second great leap by doing something even greater for humanity: investing in water research."

Beyond R&D, there are other policy approaches to consider. One policy option is to develop a new clean energy standard that incorporates water consumption as well as emissions. To date, most clean energy standards for pollution from power plant smokestacks and automobile tailpipes have focused on emissions only. Those policies set standards for emissions rates, using rates like grams, pounds, or tons of emissions per megawatt-hour of electricity generated, or grams emitted per mile of travel for cars. They also use volumetric fractions in parts-per-million, or ppm, of pollution.

For automobiles, recent fuel economy standards, implemented in miles per gallon, were developed using emissions standards of grams of carbon dioxide emitted per mile traveled.

To comply with the regulations, sophisticated scrubbers have been developed to remove pollutants such as nitrogen oxides and sulfur oxides from power plant smokestacks. Power plants fueled by nuclear, solar energy, and wind turbines meet the standards without scrubbing. As new regulations for mercury, particulate matter, and carbon dioxide are implemented, additional scrubbers might need to be installed for conventional fossil plants.

Keep in mind, however, that the scrubbers can be water intensive, as are the cooling systems for nuclear power plants. By making air emissions comply with stricter standards, the power sector might become more water intensive from the additional scrubbers or from fuel switching to nuclear power. In this instance, one environmental objective—cleaner air—can conflict with another environmental objective—reduced water use.

Adding a water standard to the clean energy standard—that is, requiring production to be done with water consumption below a particular threshold—will help mitigate this problem. But, doing so would exclude traditional nuclear plants, coal plants with water-intensive scrubbing systems, and concentrating solar thermal plants. Creating a new clean energy standard that includes a water standard along with an emissions standard might lead to a sweet spot of power plants that are both carbon-clean and water-lean. For an integrated clean energy standard, the list of "clean" power plants is essentially reduced to small nuclear plants, integrated gasification combined cycle coal plants with dry cooling, natural gas combined cycle plants, natural gas combustion turbines, solar photovoltaic cells, and wind.

Beyond policies, the dysfunctional markets for energy and water also need to be refined. In the current paradigm, many prices for energy are set by central regulators. Such fixed-pricing schemes create the preposterous situation where the price for electricity is the same for many communities in the afternoon in August as the early morning in March, even though the demand for electricity is much different and the supply is different, too. The situation is similar for water in that prices are usually fixed and do not change, despite the fact that water demand is usually higher in the summer when water supply from precipitation is usually lower. To make matters worse, as noted above, the water is often free.

The smart technologies described earlier enable a more efficient market, for which the prices for power reflect up-to-date assessments of demand and supply. Gasoline prices go up in the summer when demand is high or after supply disruptions when supply is low. Electricity prices should do the same. With the advent of smarter technologies, markets can move to a design that has more fidelity. Those technologies could track movements in supply or demand that are updated every second and are shared with consumers so that price signals could inform behavior. That is part of the appeal of the smart grid.

If we think the power markets are dysfunctional, the water markets are much more backward. While energy systems have poorly resolved metering and fixed prices, in many places water systems have not had meters at all and the water is free or fixed at a falsely low price. Yet the same approach of improving markets that would help the power system could also be used for water.

Because many agricultural and industrial customers bought their water rights a long time ago, their prices for water are often very low or highly subsidized from government projects such as western dams. For many agricultural operations, which use tremendous volumes of water, it is cheaper to waste water than to conserve it. Because water has such a low price, there is essentially no cost for wasting it. By contrast, conservation is expensive because it often requires investment in capital-intensive equipment that is more water efficient. That means it's free to throw the water away, but saving it would cost these producers millions of dollars. Heartbreaking.

Some petrochemical complexes in Texas have a similar situation, with take-or-pay contracts that discourage conservation. As the water is prepaid, plant operators have an incentive to use all the water that was purchased. If they do not use it one year, then they might forfeit the excess water the following year, creating a perverse incentive to avoid conserving water and to use as much as possible.

Another factor to consider is that water and energy are both too cheap. They are priced at levels far below their true value. In particular, the prices for energy and water in the United States are cheaper than in just about every other developed country. The low prices discourage conservation because they send a market signal that the resource is plentiful and not worth saving. That price disparity is one of the reasons Europeans and Japanese consume lower amounts of energy and water per capita. Unfortunately, low prices in the United States make it harder for people to realize the true worth of water and energy.

Also, in many places, water bills are mostly computed as cost recovery for the capital expenses of the infrastructure. Most bills require consumers to pay for the pipes and pumps, but do not always charge them for the water. Or if they charge for the water, the price per gallon is very low. In places like Sacramento, California, for many years homes did not have water meters, as there was not a charge per gallon. Essentially, consumers received a fixed water bill that did not change no matter how much water they consumed. In Ireland, water meter installations and the concept that people should pay for water led to widespread protests and discontent.

Such fixed prices for either energy or water suggest to consumers that they should consume as much as they want. At the same time, another phenomenon kicks in when fixed prices every month are high and the marginal cost for the water or electricity is low or free: people want to consume more energy and water because they want to get their money's worth for the fixed price. If they pay $40 each month in their water bill before they use any water, they feel like they need to consume a lot of water to justify the expense.

Another problem occurs when consumers get volumetric discounts: the more they use, the cheaper it is. This kind of price system is not that different from buying groceries in bulk as a way to save money. The price reduction with increasing volumes is an honest reflection of how the costs drop for the provider with economies of scale. But that structure encourages more consumption and discourages conservation as consumers chase the better deal. For constrained resources—such as water and fossil fuels—

the consequences of encouraging consumption are felt societywide. At the same time, as noted above, just increasing rates to encourage conservation can price poor people out of the market, denying them economic opportunity, comfort, and in some cases good health.

Another option would be a high price for both the fixed charges and the volumetric usage to discourage profligate consumption. High prices will reduce consumption, but they pose two different problems. First, the high prices will reduce access to energy and water for poor people who need it for economic opportunity and for many basic functions such as heating, cooking, cleaning, and drinking. There is a whole swath of the population for whom high prices would mean that water and electricity would be cut off, which would be bad from a humanitarian perspective. Second, high prices for each marginal gallon of water or kilowatt-hour of electricity sold give the utility an incentive to sell as much water and energy as possible, which might work against conservation. Many people have proposed decoupling revenues from volumetric sales as a way to get utilities interested in conservation. The way to do that is to raise the fixed prices and reduce the marginal prices for consumers' bills. That way no matter how much the customer consumes, the utility collects the same amount of revenue. But that brings us back to the first problem noted above: fixed bills that do not change no matter how much the customer consumes send an unclear signal about resource scarcity and abundance.

So what is good for encouraging conservation by the consumers—low fixed prices for the infrastructure and high prices for energy and water—is different than what encourages conservation by the utility—high fixed revenues with low marginal revenues. A common way to solve this problem is for rate setters to split the difference for the fixed prices and rates.

In parallel, there is a tension between the human right for water, which reflects social justice priorities, and the value of water as a commodity. Advanced markets for water would enable the more efficient allocation of the resource based on economic value. Heading that direction would help avoid the problems of using a lot of water for growing low-value crops or wasting water because using it is cheaper than conserving it. So proper pricing and market efficiency can offer a lot of improvements.

At the same time, there is a concern that in the process of properly valuing water, markets would make clean, accessible water a luxury good that is too expensive for many people. Although raising the price of water is generally a good thing, as water is underpriced today, the fears that markets would overlook the value of water as a staple are widespread.

Thus, there is a conflict between the social justice value of water as a right versus the market allocation value of water as a commodity. One way to accommodate these different angles is to make the quantity of water available for daily living needs—washing, cooking, eating—very cheap or free, after which the prices for luxury uses—watering lawns, washing cars—go up.

This system is known as inverted block pricing. The first increment of electricity and water—say the first 500 kilowatt-hours of electricity consumption over a month and the first 2,000 gallons of drinking water—are relatively cheap. That way people who are using the electricity and water for basic functions such as refrigeration, lighting, heating, drinking, and washing can afford to get the resources they need at a reasonable cost. Above that, different price tiers kick in that get progressively steeper. The price structure is inverted: instead of rates dropping as consumption increases, the rates increase. That structure has been implemented for water in Irvine, California, and El Paso, Texas, and for electricity in many locations.

In El Paso, an arid part of the country, where one of the world's mighty rivers, the Rio Grande, runs dry many months of the year, the first tier of water use is set as the average of the three winter months. That first tier is the cheapest, and any use above that tier is priced at a higher level. This approach presumes that users are not irrigating in those months, and thus the water use over that time period is strictly for indoor use. The rate is adjusted each year for each water meter, which helps update the baseline when the number of household occupants changes.

Presumably the first tier of rates is for basic functions that are necessary for modern society, whereas the higher rates are for the consumption tied to unnecessary functions. The rationale is that poor households consume less energy and water, and rich households consume more. The expectation is that inverted block pricing would not unduly increase the burden on people already struggling with poverty.

This expectation is not always true, however: many poor households were very high consumers of electricity. This phenomenon occurs for a variety of reasons. Many poor households have a greater number of people living in the home, for example. It is not unusual for three generations to live together and for family sizes to be much larger than for the wealthier families who have fewer children and whose grandparents live elsewhere. Because consumption tracks roughly with population, these poor, large households consume a lot. At the same time, poor families often could not afford energy-efficient air conditioners, double-pane windows, or extra attic insulation. Consequently, their homes were leakier and more difficult to keep cool or warm, also driving up their electricity consumption.

In contrast, wealthier households could afford many of the systems and components that drive down electricity consumption. Solar panels, green building designs, efficient air conditioners, and other energy-saving items all conserve energy but cost a lot of money. So, ironically, some of the city's smallest electricity consumers were among the richest households. Overall, that means there might be some surprises as the inverted block pricing gets implemented: it will encourage high-use consumers to try harder at conservation, but it might also accidentally increase the burden on households that already cannot afford the equipment they need for conservation in the first place. It is not clear if the same phenomenon happens with water. Generally speaking, water use tracks affluence more closely, as water use correlates with yard size and how expensive the neighborhood is, which translates into pressure from neighbors to irrigate lawns.

When adjusting prices, whether by markets or regulated price tiers, it is clear that prices need to go up. One of the reasons that energy and water prices in the United States are so low is because their externalities are not figured into the transaction. Externalities are costs that are borne by the consumer, but in a method that is external to the market.

There are several types of externalities. National security externalities are related to protecting the imports of petroleum with the U.S. military. In addition, there are environmental externalities related to energy consumption from air pollution, water use, and ecosystem impacts. These costs are associated with energy and water use, but are not paid through our utility bills. Rather, we pay for them in our tax bills, health care premiums, or in the economic loss related to shortened lifespans, poor health, and degraded biodiversity.

The national security costs of petroleum alone work out to be $0.20–1.00 per gallon of gasoline.[4] In addition there are the environmental impacts. In a landmark study, the National Academies of Sciences and Engineering produced a report entitled "The Hidden Costs of Energy: Unpriced Consequences of Energy Production and Use," which tried to put a price on the environmental externalities associated with energy consumption with a focus on the power sector.[5] Looking at air pollution, the report's results valued the damage done by power plants. This damage included premature mortality and morbidity, reduced grain crop and timber yields, and other monetized damages. Premature mortality had the biggest impact. Their analysis determined that coal-fired power plants cause anywhere from 0.5 to 12 cents per kilowatt-hour of damage from sulfur oxides, nitrogen oxides, and particulate matter emissions. The mean damage by coal plants is 3.2 cents per kilowatt-hour, which is very similar to the wholesale price. Each kilowatt-hour of electricity from a coal-fired power plant is worth about 2 to 4 cents, which we remit in our utility bill at a retail rate of approximately 10 to 12 cents after including the infrastructure and markups. That means if we include the 3.2 cents of damage to our lungs and environment, the full cost is about twice what the wholesale markets imply.

We pay for the latter parts through elevated insurance premiums and degraded economic output. Because their costs are borne outside of the transaction—that is, they are not listed on the utility bills we receive—they are considered market externalities. And, by the way, those estimates for monetized damages do *not* include the impacts from greenhouse gas emissions, water use, or land disturbance, all of which might drive the hidden costs—the externalities—higher.

That report was followed up by a subsequent study performed at Harvard. Researchers concluded after a more comprehensive look at the externalities of coal, including land disturbance and a few other factors left out by the National Academies, that the life-cycle impacts of coal are anywhere from 9 to 27 cents per kilowatt-hour.[6] We pay these costs through lost economic activity in other sectors, higher tax bills to our governments to pay for environmental cleanup, and higher health care premiums. And, these costs are much higher than the subsidies offered to alternatives like renewable energy.

For the energy and water markets to function more effectively, these costs need to be embedded in the price signals. Once full-cost accounting is implemented, then it is expected that the markets can settle out toward a more efficient, cleaner solution without the hidden costs. However, as long as subsides, which distort the markets, are in place, and as long as the externalities, which also distort the markets, are not priced into the system, then the markets will produce outcomes that are inconsistent with long-term goals for environmental sustainability.

One possible solution for the markets includes liberalization. An updated market design, where the prices for water and energy include full-cost accounting, adjust according to supply and demand, and include externalities, offers the opportunity for some additional efficiencies and environmental protection. By including the environmental externalities, it will be expensive to pollute and there will be an incentive to be clean. Updated markets also open up the prospects for cross-sectoral water deals that are good for energy and water. It is not unusual today for the agricultural sector to use 80 percent of the water in a region to generate less than 1 percent of the local economic output. For example, in South Texas a significant amount of water is used to grow alfalfa, a form of cattle feed, and other low-value crops. California endures a similar phenomenon.

That means in today's typical situation, the agricultural sector has a lot of water, but not a lot of money. Then along comes the energy sector as a new marginal water user. In this case, they have a lot of money, but do not have a lot of water. And, because they are the newest water user, they often provoke a lot of resistance from local communities despite the fact that their water needs are usually much lower than the agricultural sector's. Hydraulic fracturing holds much promise to increase local employment, taxes, and economic activity dramatically, but it requires significant volumes of water to do so. If the producers cannot get access to water, then their production might be curtailed.

While this competition for water resources can create conflict, it also has the potential to use efficient markets to reallocate resources. In particular, since the energy sector is rich and wants water and the agricultural sector is poor and has water, they could make a deal: they could trade money for water. This type of exchange is pretty obvious and is what we do for many commodities in other parts of the world. For the energy producers, the water is highly valuable: they can produce much greater economic value per gallon of water than the farming operations. That means they can pay a higher price for water than prevailing rates. At the same time, as noted earlier, farmers cannot afford to conserve water because the water is cheap and the equipment they need is too expensive.

If the market is structured the right way, then the energy producers can simply buy or lease the water from agricultural users. At the right price, the amount of money that is paid to the farming operations gives them the capital they need to invest in water-efficient irrigation equipment.[7] In this case, the energy sector essentially pays for new irrigation systems on the farms. Once they have that water-efficient equipment in place, farmers should still be able to produce their agricultural goods, but with much less water. This is "more crop per drop" coming to fruition. This water efficiency means farmers would have excess water available to sell to the energy producers in exchange for the capital investments. And, the potential efficiency gains on the farm are often so large that there might even be additional water leftover for the streams and basins. Just by putting a higher price on water and liberalizing the markets, the energy sector, which normally is a source of frustration as a marginal water user, becomes a catalyst for investments that increase the utility and availability of water overall. This scheme of using outside investment to achieve better management of water rights through efficiency, yielding additional water for selling or leasing, is already under way. Private equity groups are buying up farming operations in western states to secure the water rights. Then they invest in efficiency, keep the water they need for farming, and sell the rest.

The Australians and the French have already pushed their markets ahead. While there are fair concerns that water markets would make water too expensive and out of reach for poor people, that phenomenon hasn't unfolded in France, where water demand has been met by private suppliers for over a century. Leave it to the socialist French to deploy water markets with high levels of private-sector participation while the capitalistic Americans frequently use centrally regulated markets and government monopolies. In fact, French water companies have created recognizable global brands in the process: Perrier, Evian, and Vittel are all available in bottles in U.S. grocery stores. The Australian situation is more severe, as massive drought triggered a major rejiggering of their water markets. By reallocating water and putting a price on it, Australians are in position to weather the next drought more successfully. Placing a price on water makes people value it more.

Given all these idiosyncrasies of the markets, it's a sign that there is much room for improvement. Setting a real price on water and energy might solve many of these dysfunctionalities. Another major market adjustment that could be quite beneficial is a widespread shift from cost-based capitalism to value-based capitalism. Another way to describe it is a shift from manufacturing to services.

In today's utility markets, the prices are based on costs of service—it is cost-based capitalism. The utilities figure out what it costs to build a large-scale electricity or water system, amortize that amount over a twenty- to forty-year period, add in a predetermined profit—10 percent is typical—then they charge the ratepayers an amount that will recover those costs. This kind of cost-plus mentality has been the driving force for regulated monopoly utilities for a century. Their goal is not explicitly to maximize profit, but seeking that goal is hard to resist.

One way to think about that setup is that utilities are not really in the business of providing electricity, water, or natural gas; rather they are in the business of spending money on large projects, for which they slowly charge ratepayers. Selling the energy or water is sort of a by-product or an afterthought. And because most utilities have volumetric sales, the utilities have an incentive to sell more: the more they sell, the more revenue they make and the more quickly they can pay off their capital investments. Once the power plants or water systems are paid off, the utilities earn higher profits.

This whole scheme gives utilities an incentive to spend as much money as possible to increase costs and to sell as much product as possible to increase revenues to cover those costs. These factors combine to push utilities toward investing in larger, capital-intensive projects and to encourage consumption. These large projects create a capital "lock-in" effect that discourages conservation or newer, smaller, distributed technologies.

Another approach is to shift toward value-based capitalism. In value-based capitalism, companies can charge based on the value of the service they provide rather than the cost of the capital they invested. Doing so gives the providers an incentive for reducing investments and consumption as a way to increase profit margins.

Rather than selling water and electricity, the utilities could sell water services and electrical services. Instead of selling water for dishwashers, the water utility could sell a dishwashing service charged by the number of loads washed. Instead of selling electricity for lighting, the electrical utility could sell lighting services sold by the number of lumens of lighting. Instead of selling electricity for water heaters, the electrical utility could sell electrically heated water valued by the number of gallons of hot water.

Today, the utility sells kilowatt-hours of electricity and the customer converts it in the home to lumens of lighting. Most incandescent lightbulbs have a 5 percent efficiency, which means for every one hundred watts of power coming in the home for an incandescent lightbulb, only five watts is emitted as useful light, with the other ninety-five watts released as waste heat. Today's utility likes it when customers have inefficient incandescent lightbulbs because that drives up consumption, increasing revenue.

Switching toward a service-based model opens the door for significant savings because the utility's incentive switches from encouraging its customers to increase consumption to encouraging conservation instead. If the utility sells lighting services, measured in how many lumens of actual lighting is delivered, instead of electricity measured in kilowatt-hours, then the utility would have an incentive to install the most efficient lightbulbs possible. If the utility owned the lightbulbs and sold lumens, then they would prefer an LED system with 20 percent efficiency instead of the 5 percent efficiency of incandescents. To deliver the same five watts of lighting, only twenty-five watts of electrical power would be needed, which is a significant savings compared with the one hundred watts that would have been needed for the older lightbulb.

While customers could make this switch on their own, there are two typical barriers for doing so.[8] The energy-efficient appliances usually cost more money to purchase and install, although they save money in the long run because they consume fewer resources. The savings might be attractive but out of reach if homeowners do not have the money to pay up front for the installation. Also, consumers often do not have enough knowledge or confidence in the newer efficient options. By contrast, the utilities have both the expertise to know which options are available and the money that is necessary to install the more expensive items.

For the cost-based model, customers use inefficient lightbulbs and the utility's money and expertise sit on the sidelines. For the service-based model, utilities want their customers to use the efficient LEDs. By switching the incentive, the utilities become a powerful partner for upgrading equipment in homes and buildings. The utilities can pay to install the better devices, and then collect money on the savings. Doing so aligns the utility's profit incentives with conservation: the more efficient the devices, the higher the profit margins. Consequently, the customer, the utility, and the environment would all benefit.

This transition can be illustrated with two examples: a photocopier manufacturer and a car paint maker. For a manufacturer of photocopiers, the goal is to sell as many copiers as possible. As with all manufacturers, the perverse incentive is to make machinery that eventually breaks down, so customers need to buy new equipment. However, by switching from photocopier manufacturing to providing document services, a company can move from cost-based capitalism to value-based capitalism. For the new model, the company would make money based on delivered services that their equipment provided—such as copies made or documents scanned—rather than simply selling the equipment itself. For the new business model, it is in the company's interest to manufacture equipment that lasts as long as possible. By switching from a cost-based manufacturing model to a value-based service model, the company's incentive changes from consuming resources for manufacturing to extending the utility of those resources, by making their machines last as long as possible.

Car paint manufacturers could do something similar. In the cost-based volumetric sales model, they want to sell as much paint as possible to the automakers. When automakers are sloppy and spill paint on the factory floor, that is good for the paint company's business and gives them a profit motive to hope that the car manufacturers are wasteful. But, instead of selling paint, the company could sell painted cars. By being integrated with the manufacturers' assembly lines, the paint company could make the paint and also apply it. Doing so leverages their expertise at how to most efficiently apply the coatings. And, in the service model, they have an incentive to make sure that not a single drop of paint is wasted. Before, every bit of wasted paint was good for sales, but with the paint company in charge of applying the paint, every drop of wasted paint means smaller profit margins for them.

This arrangement is similar to what could happen with utilities: in the cost-based model, they want us to be wasteful and buy as much electricity, natural gas, and water as possible. But in the services model, if they use their expertise they could implement more efficiency to increase their profit margins.

Even better would be an integrated service utility. In many locations the electrical, water, and gas utilities are separate. However, the better solution might require cooperation among all three. Take water heating, for example. The electric utility that shifts to a service model would help consumers save energy for their water heating needs. They could get a much more efficient electric water heater that is better insulated and has a larger tank. The better insulation will save the consumer money because less heat is lost to the atmosphere, and the larger tank means the water heater is less likely to be on during peak hours, which would help the utility with grid management and help the consumer avoid peak-time prices. Because these better water heaters are more expensive than conventional water heaters, the cheaper, less efficient, smaller systems are usually installed instead. This case is the typical consequence of what happens when the person who pays for the installation of the water heater—usually the home builder—is different from the person who pays for the operation of the water heater, which is the homeowner. Reducing operational costs for the homeowner requires the home builder to spend more money on the appliance. When they are building a custom home, owners can easily specify to the builder that they want a better heating system, but most homes are not custom built. Overcoming the cost barrier of the more expensive and efficient system is one of the main advantages of having a utility as a partner.

With the electric utility as a partner, the consumer could have a more efficient electric water heater installed. But even better would be an efficient natural gas water heater, which can avoid the losses at the power plant. Although an electric utility with a service model could implement more efficient electric water heaters as part of its business plan, it is unlikely to recommend a natural gas water heater, because that would cut into their service. An integrated electric and gas utility, however, could do it. Best of all, an integrated electric, gas, and water utility might recommend a solar water heater with gas backup and double-redundancy with electric backup to the gas backup. That approach would save even more gas and electricity while still making the utility money.

The switch from cost-based capitalism (the manufacturing model) to value-based capitalism (the service model) could be very profitable. To use a rough analogy from the world of computers, most computer makers operate on the cost-based model: they strive to make the world's most sophisticated supply chains as a way to reduce costs. After they determine the cost to customize, manufacture, and deliver their computers, they add a small markup for profit and determine the sales price. In contrast, Apple operates on a value-based model. Rather than focusing on cost, they strive to improve the user experience for the consumer. Then they figure out how much their products are worth to consumers, set the price at that amount, and then strive to reduce their manufacturing costs as a way to improve price margins.

For this simple example, the service model appears to be much more profitable. Apple's profit margins are much higher, and sales outputs per square foot of retail space are the highest in the world (at more than $6,000 per square foot, about double that of luxury retailer Tiffany).[9] Apple often competes to be the world's most profitable and valuable company, serving as an example of the profit potential for value-based approaches to business. If the same possibilities can be applied to the world of energy and water, then profit margins will increase for those sectors because of efficiency and conservation, rather than in spite of it.

At their core, these nontechnical solutions aim to address the following policy and ethical conundrums that our society must strive to address for the energy-water nexus. Some of these conundrums and questions are exacerbated by climate change.

One of those is the tension between availability and price, and the human right to water and energy versus the commodity value of water and energy. How should energy and water planners prioritize ensuring some minimum availability at a price that is affordable, while also meeting the needs to be environmentally and economically sustainable? Is there a way to use markets to increase the availability of these resources and to foster a culture of conservation without pricing people out of the market? If energy or water is too expensive, people cannot afford it, putting their life, liberty, and economic interests at stake. If the resources are too cheap, then society wastes them, draining reserves and degrading the environment. Are energy and water a human right, one that everyone deserves guaranteed access to, or are they commodities whose availability, price, quantity, and quality should be determined by the markets? Can we combine both concepts with some minimum threshold of the human right to water and energy, above which they are commodities? Who owns the water: nature or people? All humans equally, or individuals based on their wealth or position of authority?

What about balancing the struggle of quantity and quality? How do we ensure that energy and water are available at the right quantity without compromising the quality? For decades, "dilution is the solution" was a mantra to improving water quality. By diluting water with additional quantities, the polluted or degraded water would improve. But, increasing the quantity takes energy and money, which has its own impacts on water quality from secondary pollution. Do we need more water and energy? Or simply cleaner water and energy?

Which societal structures—privatized or socialized—are best suited to meeting our energy and water needs in the long term? Socialized structures with centrally owned water systems, which is what the United States has used for a century, or private markets that can raise the capital necessary for investments and innovation? Socialized systems have the advantage of pooling resources to the benefit of society with less risk of making water too expensive for the poor. But socialized systems innovate slowly and are clumsy at best. Privatized systems are more nimble and advanced, but raise the specter of concern about prices going too high for most people.

All of these tensions will only worsen due to our shifting climate, which will heighten intergenerational and intercontinental disparity. To avoid the accumulating exacerbation of these problems, we must change today. It's hard enough to clean up our own backyards for our own benefit; do we have the discipline to clean up our yard for the benefit of people who live around the world and people who haven't been born yet? Making matters trickier, it is the rich part of the world that consumes the most energy and emits the most greenhouse gases. That means it is the rich who must change for the benefit of the poor. If recent political alignments are any indicator, this is a tough proposition.

Despite decades of calls for using less freshwater, this conservation measure hasn't taken root everywhere. Generally, using a professional car wash can dramatically reduce the amount of water required per car. While a normal citizen might use 150 gallons of water to wash a car at home, a professional service with reclamation uses less water to wash a car than is used for a shower.[10] Unfortunately, not every car wash follows this approach.

One of my favorite barbecue restaurants in Austin is combined with a gasoline station. They have a car wash on-site as part of their mix of services and advertise their use of 100 percent nonrecycled hot water. They proudly proclaim, "All water is fresh—we never reuse dirty water." This company considers it a selling point not to recycle its resources. What is particularly striking is that this car wash is in the middle of a drought-prone

A car wash in Texas advertises its use of hot water and its decision not to recycle it. [Photo by Jeffrey M. Phillips, November 2014]

state. The good news is that not every car wash has that attitude. Another car wash in my neighborhood brags on its website about its water conservation programs. This one proudly advertises that it has been using reclaimed water since 2006 as a way to reduce water needs per car.

Overall, changing our attitudes will be the shift we need to make. We need to change the way we think about energy and water. All of these problems and all of these solutions—technical and nontechnical—point to an obvious starting point: conservation. The good news of the energy-water nexus is that water conservation saves energy and energy conservation saves water. Conservation is a cross-cutting solution, but one that hasn't been fully adopted, despite advocacy going as far back as the 1960s.

Thankfully, younger generations are already there. I see it with my college students who use the same water bottle for a year, take the bus to school, and join ridesharing programs. Even younger—and more promising—are schoolchildren.

A stamp encouraging water conservation from 1960. [United States Postal Service, © Can Stock Photo Inc./AlexanderZam]

When my daughter was seven years old, we had a nightly ritual of brushing our teeth together. We would turn on the faucet to wet our toothbrushes, turn it off while we brushed to save water, then turn it back on again to rinse. One night I didn't turn the faucet off quickly enough to meet her satisfaction. She glared at me, turned the faucet off abruptly, then told me adamantly: "Turn off the water, Daddy. The scientists need time."

I was dumbstruck. Her statement summed it up nicely. Conservation doesn't solve all of our problems—it's hard to light a lightbulb with conservation, for example—but it sure does buy us some badly needed time. Scientists can use that time to invent the technologies we need, and society can use that time to implement a paradigm shift.

Conservation is the obvious solution. It's the critical starting point that buys us time. It is also one of the few solutions that works on a very small scale and a very large scale. And there are many ways to make a huge impact. Turning off the faucet while brushing teeth is one of those simple conservation acts that cost nothing—they save money—though in all honesty, that action saves very little water. But every little bit helps. If we all do it, then it will save billions of gallons cumulatively.

There are even better opportunities. If we really want to save water, we shouldn't plant thirsty lawns that require water to grow and gasoline to mow, and we should change our entire approach to irrigated agriculture. If we want to save energy, we need better and smarter power plants, cities, and vehicles. We should stop leaks of energy and water, install water-efficient and energy-efficient technologies, update our markets, and shift our mindsets. When we start thinking about the world with the clarity of my seven-year-old daughter, then we will be well on our way toward our destination: a cleaner, more prosperous, and more sustainable world.

Notes

Chapter 1. Healthy, Wealthy, and Free

1. The story of drought in the southeastern United States, the border tensions with other states, and the relationship with power plants have been covered by multiple articles, for example: L. Mungin, "Two Off-Line Power Plants Help Region Hit Water Goal," *Atlanta Journal-Constitution,* December 20, 2007; Theo Emery, "Georgians Want Access to Tenn. Water," *Tennessean,* February 8, 2008; and Toluse Olorunnipa and Michael C. Bender, "Florida to Sue Georgia in U.S. Supreme Court over Water," *Bloomberg,* August 14, 2013.
2. S. Gottipati, "India Endures World's Largest Blackout," *New York Times,* July 31, 2012, and J. Yardley and G. Harris, "Second Day of Power Failures Cripples Wide Swath of India," *New York Times,* July 31, 2012.
3. Brent Kallestad, "Fla. to Sue Army Corps of Engineers over Water," *USA Today,* June 20, 2008.
4. Felicity Barringer, "Lake Mead Could Be Within a Few Years of Going Dry, Study Finds," *New York Times,* February 13, 2008.
5. "Salt Water Plant Opened in London," *BBC News,* June 2, 2010.
6. The points about the hydraulic theory of civilization, the Chinese empires, and the importance of controlling the Yellow River are covered by Fred Pearce in his book *When the Rivers Run Dry: Water—The Defining Crisis of the Twenty-First Century* (Boston: Beacon, 2006). The comment about the Chinese character for politics is from an article in the *Economist* titled "Awash in Waste: Tradable Usage Rights Are a Good Tool for Tackling the World's Water Problems," April 8, 2009.
7. The background on Roman aqueducts is from A. T. Hodge, *Roman Aqueducts and Water Supply,* 2nd edition (London: Bristol Classical, 2002).
8. Richard Stone, "Divining Angkor," *National Geographic,* July 2009, and Pearce, *When the Rivers Run Dry.*
9. The San Francisco Public Utilities Commission has a brief write-up of two water temples: the Pulgas Water Temple built in 1934 to celebrate water coming to San Francisco from the Hetch Hetchy reservoir, and the Sunol Water Temple built in 1910 to commemorate a previous water system (www.sfwater.org).

10. See, for example, Jared Diamond, *Collapse: How Societies Choose to Fail or Succeed* (New York: Viking, 2005), and Brian Fagan, *The Great Warming: Climate Change and the Rise and Fall of Civilizations* (London: Bloomsbury, 2008).
11. P. Zhang et al., "A Test of Climate, Sun, and Culture Relationships from an 1810-Year Chinese Cave Record," *Science,* November 7, 2008.
12. Katherine Unger, "Drought to Blame for Rome's Decline?" *Earth,* February 2009.
13. Guy Gugliotta, "The Maya: Glory and Ruin," *National Geographic,* August 2007; Heather Pringle, "Did Pulses of Climate Change Drive the Rise and Fall of the Maya?" *Science,* November 9, 2012; Heather Pringle, "A New Look at the Mayas' End," *Science,* April 24, 2009.
14. Fagan, *The Great Warming.*
15. See, for example, Peter H. Gleick, "Water, Drought, Climate Change, and Conflict in Syria," *Weather, Climate, and Society* 6, no. 3 (July 2014).
16. The story of the discovery of the connection of cholera with contaminated water and the Great Stink in London has been extensively covered. I recommend Bill Bryson's *At Home: A Short History of Private Life* (New York: Anchor, 2011) as accessible and interesting to a general audience.
17. David Sedlak, *Water 4.0: The Past, Present, and Future of the World's Most Vital Resource* (New Haven: Yale University Press, 2014).
18. For information on global access to water and wastewater, the United Nations is a reliable source. For information on global water stress, Vörösmarty's work is the standard-bearer: C. J. Vörösmarty et al., "Global Threats to Human Water Security and River Biodiversity," *Nature,* September 30, 2010. In addition, the Pacific Institute produces a series of biennial reports on freshwater resources with convenient summaries of water data and in-depth analyses on water topics, including availability, access, policies, and technologies: Peter Gleick et al., *The World's Water,* volume 8 (Washington, D.C.: Island, 2014, and previous volumes). For 100 million people in China: Richard Stone and Hawk Jia, "Going Against the Flow," *Science,* August 25, 2006. For 2.6 billion people: Alan Fenwick, "Waterborne Infectious Diseases—Could They Be Consigned to History?" *Science,* August 25, 2006. For 1.8 billion people by 2025: United Nations, "Water Scarcity Factsheet," 2013 (www.unwater.org/publications/publications-detail/en/c/204294, accessed August 22, 2015).
19. For statistics on asthma and air quality impacts, see American Lung Association (www.lung.org/associations/states/colorado/asthma/stats.html, accessed December 29, 2014); U.S. Environmental Protection Agency; World Health Organization; and Fabio Caiazzo et al., "Air Pollution and Early Deaths in the

United Sates, Part I: Quantifying the Impact of Major Sectors in 2005," *Atmospheric Environment* 79 (2013).

20. Edward Wong, "Air Pollution Linked to 1.2 Million Premature Deaths in China," *New York Times,* April 1, 2013.
21. For global statistics on energy consumption, see the International Energy Agency's databases and reports (www.iea.org).
22. For statistics on water use by sector, the U.S. Geological Survey provides information for the United States, and the World Bank gives global information. Jill Boberg's report *Liquid Assets* (Santa Monica: Rand Corporation, 2005) is a convenient overview and summary of some of the key underlying issues.
23. World Health Organization, "Burden of Disease from Household Air Pollution for 2012," (www.who.int/phe/health_topics/outdoorair/databases/FINAL_HAP_AAP_BoD_24March2014.pdf).
24. S. R. Kirshenbaum and M. E. Webber, "Liberation Power: What Do Women Need? Better Energy," *Slate,* November 4, 2013.
25. "Facts About Global Poverty and Microcredit," *To Our Credit: A Two-Part Program for PBS* (www.pbs.org/toourcredit/facts_one.htm, downloaded May 25, 2015).
26. For a discussion of appliances and women, see Institute of Electrical and Electronics Engineers (IEEE) Global History Network, "Household Appliances and Women's Work," 2012; and "Fridges and Washing Machines Liberated Women, Study Suggests," *ScienceDaily,* March 13, 2009.

Chapter 2. Energy

Epigraph: Richard P. Feynman, Robert B. Leighton, Matthew Sands, *The Feynman Lectures on Physics,* volume 1, 4-2 (Addison-Wesley, 1963).

1. For a good review of the history of thermodynamics, see Bruce Hunt, *Pursuing Power and Light: Technology and Physics from James Watt to Albert Einstein* (Baltimore: Johns Hopkins University Press, 2010). For a good introduction to the fundamentals of thermodynamics, see Michael E. Webber, *Thermo 101: Introduction to Engineering Thermodynamics* (Austin: University of Texas Press, Tower Book Imprints, 2015).
2. Hunt, *Pursuing Power and Light.*
3. M. E. Webber, "Redefining Humanity Through Energy Use," *Earth,* March 2010.
4. Lawrence Livermore National Laboratory publishes new estimates each year of energy consumption in the United States, including the relative fractions for end-uses and waste.

5. For a list of conversions and their typical efficiencies, see J. Tester et al., *Sustainable Energy: Choosing Among Options,* 2nd edition (Cambridge: MIT Press, 2012).
6. Vaclav Smil, *Energy: A Beginner's Guide* (Oxford: Oneworld Publications, 2006).
7. W. Cronon, *Nature's Metropolis: Chicago and the Great West* (New York: W. W. Norton, 1992).
8. For global statistics on energy consumption, see the International Energy Agency's databases and reports, for example the World Energy Outlook, published annually.
9. For an interesting anecdote about the U.S. secretary of energy Stephen Chu discussing the excellent technical performance of petroleum-based fuels, see Russell Gold, *The Boom: How Fracking Ignited the American Energy Revolution and Changed the World* (New York: Simon and Schuster, 2014), chapter 3.
10. For global statistics on energy consumption, see the International Energy Agency's databases and reports, for example the World Energy Outlook, published annually. For information on United States energy consumption, see the U.S. Energy Information Administration's databases and reports, such as the Annual Energy Review.
11. Cronon, *Nature's Metropolis.*
12. Alexander Starbuck, *History of the American Whale Fishery* (New York: Argosy-Antiquarian, 1878).
13. R. Chernow, *Titan: The Life of John D. Rockefeller, Sr.* (New York: Random House, 1998).
14. M. E. Webber, "The Bright Future for Natural Gas in the United States," *Earth,* December 2012.

Chapter 3. Water

Epigraph: Loren Eiseley, *The Immense Journey* (New York, 1957), 15.

1. Jeremy Narby, *The Cosmic Serpent* (New York: Jeremy P. Tarcher/Putnam, 1998).
2. "America's Sewage System and the Price of Optimism," *Time,* August 1, 1969.
3. The exact contribution from water vapor to atmospheric temperature is difficult to estimate, and the scientific community does not have consensus on the exact value. Vaclav Smil estimates that water vapor is responsible for approximately 20 degrees of the 35-degree increase in temperature: Vaclav Smil, *Energy: A Beginner's Guide* (Oxford: Oneworld Publications, 2006). For an introduction to the underlying science of climate change, see Mark

Maslin, *Global Warming: A Very Short Introduction* (New York: Oxford University Press, 2004).

4. United States Geological Survey, "The Water Cycle," *USGS: Science for a Changing World* (http://water.usgs.gov/edu/watercycle.html, accessed December 31, 2014).
5. Smil, *Energy*.
6. Water values are from United States Geological Survey, "The Water Cycle." The original source for the estimates is Igor Shiklomanov's chapter "World Fresh Water Resources," in Peter H. Gleick, editor, *Water in Crisis: A Guide to the World's Fresh Water Resources* (New York: Oxford University Press, 1993).
7. The GRACE experiment has been widely covered. For two general interest sources, see NASA Mission Pages (www.nasa.gov/mission_pages/Grace, accessed December 31, 2014), and T. Green, "The Gravity of Water," *University of Texas News Features,* October 7, 2011 (www.utexas.edu/features/2011/10/07/grace, accessed December 31, 2014).
8. Ibid.
9. Global Water Intelligence, *Desalination Markets 2010: Global Forecast and Analysis,* technical report, Desal Data, 2010.
10. B. D. Lutz, A. N. Lewis, and M. W. Doyle, "Generation, Transport, and Disposal of Wastewater Associated with Marcellus Shale Gas Development," *Water Resources Research* 49 (2013).
11. Peter H. Gleick, *Bottled and Sold: The Story Behind Our Obsession with Bottled Water* (Washington, D.C.: Island, 2010).
12. Shiklomanov, "World Fresh Water Resources."
13. Ibid.
14. J. McBride, *The Color of Water: A Black Man's Tribute to His White Mother* (New York: Riverhead Trade, 1995).
15. A. Y. Hoekstra, A. K. Chapagain, M. M. Aldaya, and M. M. Mekonnen, *Water Footprint Manual: State of the Art, 2009,* technical report, Water Footprint Network, 2009; and Water Footprint Network (www.waterfootprint.org, accessed October 16, 2012).
16. The impact of drought has been covered and reported in multiple sources: John Holland, "Report: Droughts Drains $2.2 Billion from Farm Economy," *Modesto Bee,* July 15, 2014 (www.modbee.com/news/local/article3167892.html, accessed December 27, 2014); U.S. National Oceanic and Atmospheric Administration, National Climatic Data Center (www.ncdc.noaa.gov/billions/events); and Associated Press, "U.S. Drought: Half of All Counties Disaster Areas," *CBS News,* August 1, 2012.

17. The U.S. Drought Monitor is produced through a partnership between the National Drought Mitigation Center at the University of Nebraska, Lincoln, the United States Department of Agriculture, and the National Oceanic and Atmospheric Administration (http://droughtmonitor.unl.edu).
18. Jill Boberg, *Liquid Assets* (Santa Monica: Rand Corporation, 2005).
19. U.S. Geological Survey reports.
20. U.S. Department of Energy, *The Transportation Energy Data Book, 2012,* technical report; Federal Highway Administration, U.S. Department of Transportation, "Freight Facts and Figures, 2011," technical report FHWA-HOP-12-002.
21. Doris Kearns Goodwin, *Team of Rivals: The Political Genius of Abraham Lincoln* (New York: Simon and Schuster, 2005).
22. Sardar Sarovar Narmada Nigam Ltd., "Project at a Glance" (www.sardarsarovardam.org, downloaded May 24, 2015).
23. Marc Reisner, *Cadillac Desert* (New York: Viking, 1986).
24. Modern Marvels, *The Manhattan Project,* History Channel, 2004.
25. Reisner, *Cadillac Desert.*
26. Ibid.
27. Bobby Magill, "Methane Emissions May Swell from Behind Dams," *Scientific American,* October 29, 2014 (www.scientificamerican.com/article/methane-emissions-may-swell-from-behind-dams, accessed January 1, 2015); and Siyue Li and X. X. Lu, "Uncertainties of Carbon Emission from Hydroelectric Reservoirs," *Natural Hazards* 62 (2012).
28. A. Lugg and C. Copeland, "Review of Cold Water Pollution in the Murray-Darling Basin and the Impacts on Fish Communities," *Ecological Management and Restoration,* 15: 71–79 (2014); Daniel B. Hayes, Hope Dodd, and JoAnna Lessard, "Effects of Small Dams on Cold Water Stream Fish Communities," American Fisheries Society Symposium, 2006.
29. Boberg, *Liquid Assets.*
30. Xtankun Yang and X. X. Lu, "Ten Years of the Three Gorges Dam: A Call for Policy Overhaul," *Environmental Research Letters* 8 (2013).
31. For an excellent overview of water infrastructure throughout history, see David Sedlak, *Water 4.0: The Past, Present, and Future of the World's Most Vital Resource* (New Haven: Yale University Press, 2014).
32. Environmental Protection Agency, U.S. Department of the Interior, *The Clean Water and Drinking Water Infrastructure Gap Analysis,* technical report, EPA-816-R-02-020, 2002.
33. Ibid.
34. Ibid.

35. Texas Water Development Board, *The 2012 State Water Plan,* technical report, 2012.

Chapter 4. Water for Energy

1. International Energy Agency, *World Energy Outlook, 2012,* Organization for Economic Cooperation and Development, 2013.
2. International Energy Agency, *World Energy Outlook, 2013,* Organization for Economic Cooperation and Development, 2014.
3. U.S. Army Corps of Engineers, The Institute for Water Resources, *Hydropower: Value to the Nation,* Fall 2001.
4. Cutler Cleveland, "China's Monster Three Gorges Dam Is About to Slow the Rotation of the Earth," *Business Insider.com,* June 18, 2010 (www.businessinsider.com/chinas-three-gorges-dam-really-will-slow-the-earths-rotation-2010-6).
5. R. Sternberg, "Hydropower's Future, the Environment, and Global Electricity Systems," *Renewable and Sustainable Energy Reviews* 14 (2010); Jim Carlton, "Deep in the Wilderness, Power Companies Wade In," *Wall Street Journal,* August 21, 2009; Boualem Hadjerioua, Yaxing Wei, and Shih-Chieh Kao, *An Assessment of Energy Potential at Non-Powered Dams in the United States,* prepared for the U.S. Department of Energy Wind and Water Power Program, Oak Ridge National Laboratory, April 2012.
6. Toby Sterling, "Dutch Seek to Harness Energy from Salt Water Mix," *Phys-Org,* November 26, 2014 (http://phys.org/news/2014-11-dutch-harness-energy-salt.html); Sonal Patel, "Statkraft Shelves Osmotic Power Project," *PowerMag,* March 1, 2014.
7. P. Torcellini, N. Long, and R. Judkoff, *Consumptive Water Use for U.S. Power Production,* technical report, NREL-TP-550-33905, National Renewable Laboratory, U.S. Department of Energy, 2003; M. M. Mekonnen and A. Y. Hoekstra, "The Blue Water Footprint of Electricity from Hydropower," *Hydrology and Earth Systems Sciences* 16 (2012).
8. M. A. Maupin, J. F. Kenny, S. S. Hutson, J. K. Lovelace, N. L. Barber, and K. S. Linsey, *Estimated Use of Water in the United States in 2010,* U.S. Geological Survey Circular 1405, 2014.
9. Environmental Protection Agency, U.S. Department of the Interior, "History of the Clean Water Act" (www.epa.gov/lawsregs/laws/cwahistory.html, accessed October 27, 2012); and Environmental Protection Agency, U.S. Department of the Interior, Summary of the Clean Water Act (1972), 33 U.S.C. §1251 et seq. (www.epa.gov/lawsregs/laws/cwa.html, accessed October 27, 2012).

10. C. W. King, A. S. Stillwell, K. T. Sanders, and M. E. Webber, "Coherence Between Water and Energy Policies," *Natural Resources Journal,* 2013; and *Estimating Freshwater Needs to Meet Future Thermoelectric Generation Requirements,* technical report, DOE/NETL-400/2008/13391, National Energy Technology Lab, U.S. Department of Energy, September 30, 2008.
11. *Proposed Resolution by the California State Lands Commission Regarding Once-Through Cooling in California Power Plants,* technical report, California State Lands Commission, 2006.
12. C. Kutscher et al., "Hybrid Wet-Dry Cooling for Power Plants," Parabolic Trough Technology Workshop, Incline Village, Nevada, National Renewable Energy Laboratory, 2006.
13. The hybrid power plant cooling system by Johnson Controls is one notable example of a hybrid design that reduces water use; see "Thermosyphon Cooler Hybrid System for Water Savings in Power Plants," *Technology Insights: A Report from EPRI's Innovation Scouts,* November 2012.
14. K. M. Twomey, C. M. Beal, C. W. King, and M. E. Webber, "Biofuels: An Energy and Water Conundrum," *World Energy Monitor,* March 2012; C. W. King and M. E. Webber, "Water Intensity of Transportation," *Environmental Science and Technology* 42 (September 24, 2008).
15. King and Webber, "Water Intensity of Transportation."
16. Twomey et al., "Biofuels"; King and Webber, "Water Intensity of Transportation."
17. B. D. Lutz, A. N. Lewis, and M. W. Doyle, "Generation, Transport, and Disposal of Wastewater Associated with Marcellus Shale Gas Development," *Water Resources Research* 49 (2013).
18. For an excellent technical assessment of the water needed for extracting fossil fuels, see the work by the Bureau of Economic Geology at the University of Texas at Austin, including J.-P. Nicot and B. R. Scanlon, "Water Use for Shale-Gas Production in Texas, U.S.," *Environmental Science and Technology* 46 (2012): 3580–3586; and J.-P. Nicot, R. C. Reedy, R. A. Costley, and Y. Huang, *Oil and Gas Water Use in Texas: Update to the 2011 Mining Water Use Report,* 2012.
19. Matt Mantell, Chesapeake Corporation, personal communication; *Modern Shale Gas Development in the United States: A Primer,* U.S. Department of Energy, National Energy Technology Laboratory, April 2009.
20. "A Lady Comes to Burkburnett," *Cosmopolitan,* August 1939.
21. Mike Soraghan, "Shaken More Than 560 Times, Okla. Is Top State for Quakes in 2014," *EnergyWire,* January 5, 2015 (www.eenews.net/energywire/stories/1060011066).

22. Eric Hand, "Oil and Gas Operations Could Trigger Large Earthquakes," *Science,* April 23, 2015.
23. Margaret Cook and Michael E. Webber, "Food, Fracking, and Freshwater: The Potential for Markets and Cross-Sectoral Investments to Enable Water Conservation," *Water* (in review).
24. C. W. King, A. S. Stillwell, K. T. Sanders, and M. E. Webber, "Coherence Between Water and Energy Policies," *Natural Resources Journal,* 2013.
25. C. M. Beal, R. E. Hebner, M. E. Webber, R. S. Ruoff, F. Seibert, and C. W. King, "Comprehensive Evaluation of Algal Biofuel Production: Experimental and Target Results," *Energies* 5 (special issue: Algal Fuel), no. 6 (2012); and C. M. Beal, A. S. Stillwell, C. W. King, S. M. Cohen, H. Berberoglu, R. P. Bhattarai, R. Connelly, M. E. Webber, and R. E. Hebner, "Energy Return on Investment for Algal Biofuel Production Coupled with Wastewater Treatment," *Water Environment Research* 84, no. 9 (2012).
26. British Petroleum, *Statistical Review of Energy,* updated annually, and the U.S. Energy Information Administration.
27. Union of Concerned Scientists, *How It Works: Water for Coal,* 2010.
28. William J. Broad, "Nuclear Option on Gulf Oil Spill? No Way, U.S. Says," *New York Times,* June 2, 2010.
29. For a discussion of the history of hydraulic fracturing, including the Nixon-era projects related to nuclear weapons for oil and gas production, see Russell Gold, *The Boom: How Fracking Ignited the American Energy Revolution and Changed the World* (New York: Simon and Schuster, 2014).
30. P. Rogers and S. Gonzalez, "San Francisco Bay Spill Largest Since '96," *San Jose Mercury News,* November 7, 2007; and A. Price, "City Sleuths Solve Mystery of Oil Tank Owner," *Austin American-Statesman,* March 12, 2008.
31. Dina Cappiello, Frank Bass, and Cain Burdeau, "AP Investigation: Ike Environmental Toll Apparent," *USA Today,* October 6, 2008.
32. "Significant Pipeline Incidents," U.S. Department of Transportation, Pipeline and Hazardous Materials Safety Administration (http://primis.phmsa.dot.gov/comm/reports/safety/SigPSI.html, accessed May 8, 2013).
33. Environmental Protection Agency, U.S. Department of the Interior, "EPA's Response to the Enbridge Oil Spill" (www.epa.gov/enbridgespill, accessed May 9, 2013); Lisa Song, "Cleanup of 2010 Mich. Dilbit Spill Aims to Stop Spread of Submerged Oil," *InsideClimateNews,* March 27, 2013.
34. Lutz et al., "Generation, Transport, and Disposal of Wastewater Associated with Marcellus Shale Gas Development"; and Thomas H. Darrah, Avner Vengosh, Robert B. Jackson, Nathaniel R. Warner, and Robert J. Poreda, "Noble Gases Identify the Mechanisms of Fugitive Gas Contamination in

Drinking-Water Wells Overlying the Marcellus and Barnett Shales," *Proceedings of the National Academy of Sciences* 111, no. 39 (2014).

35. Darrah et al., "Noble Gases Identify the Mechanisms."
36. Thomas W. Simpson, A. N. Sharpley, R. W. Howarth, H. W. Paerl, and K. R. Makin, "The New Gold Rush: Fueling Ethanol Production While Protecting Water Quality," *Journal of Environmental Quality,* 2008; Richard Alexander et al., "Differences in Phosphorus and Nitrogen Delivery to the Gulf of Mexico from the Mississippi River Basin," *Journal of Environmental Quality,* 2008; Simon D. Donner and Christopher J. Kucharik, "Corn Based Ethanol Production Compromises Goal of Reducing Nitrogen Export by the Mississippi River," *Proceedings of the National Academy of Sciences,* 2008; Charlotte de Fraiture, Mark Giordano, and Yongsong Liao, "Biofuels and Implications for Agricultural Water Use: Blue Impacts of Green Energy," *Water Policy Supplement,* 2008; James Owen Alexander, "World's Largest Dead Zone Suffocating Sea," *National Geographic News,* March 25, 2010; K. M. Twomey, A. S. Stillwell, and M. E. Webber, "The Unintended Energy Impacts of Increased Nitrate Contamination from Biofuels Production," *Journal of Environmental Monitoring,* 2010; King et al., "Coherence Between Water and Energy Policies."

Chapter 5. Energy for Water

1. K. T. Sanders and M. E. Webber, "Evaluating the Energy Consumed for Water Use in the United States," *Environmental Research Letters* 7 (2012).
2. Marc Reisner, *Cadillac Desert* (New York: Viking, 1986).
3. Peter H. Gleick and Heather S. Cooley, "Energy Implications of Bottled Water," *Environmental Research Letters,* 2009; Peter H. Gleick, *Bottled and Sold: The Story Behind Our Obsession With Bottled Water* (Washington, D.C.: Island, 2010).
4. A. S. Stillwell, C. W. King, M. E. Webber, I. J. Duncan, and A. Hardberger, "The Energy-Water Nexus in Texas," *Ecology and Society* (special feature The Energy-Water Nexus: Managing the Links Between Energy and Water for a Sustainable Future), 16, no. 1 (2011).
5. Ronnie Cohen, Barry Nelson, and Gary Wolff, *Energy Down the Drain: The Hidden Costs of California's Water Supply,* National Resources Defense Council and Pacific Institute, August 2004.
6. Global Water Intelligence, *Desalination Markets 2010: Global Forecast and Analysis,* technical report, Desal Data, 2010.
7. R. George, *The Big Necessity: The Unmentionable World of Human Waste and Why It Matters* (New York: Metropolitan, 2008).

8. Stillwell et al., "The Energy-Water Nexus in Texas."
9. For more information, PUB, Singapore's national water agency (www.pub.gov.sg/about/historyfuture/Pages/NEWater.aspx, accessed January 3, 2015).
10. Norman Chan, "What SpaceX's Dragon Brought to the International Space Station," *Tested,* May 25, 2012 (www.tested.com/science/space/44509-what-spacexs-dragon-brought-to-the-international-space-station, accessed January 3, 2015).
11. M. E. Webber, D. S. Baer, and R. K. Hanson, "Ammonia Monitoring Near 1.5 μm with Diode Laser Absorption Sensors," *Applied Optics* 40, no. 12 (2001).
12. M. E. Webber, R. Claps, F. V. Englich, F. K. Tittel, J. B. Jeffries, and R. K. Hanson, "Measurements of NH_3 and CO_2 with Distributed-Feedback Diode Lasers Near 2 μm in Bioreactor Vent Gases," *Applied Optics* 40, no. 24 (2001).
13. Sanders and Webber, "Evaluating the Energy Consumed for Water Use in the United States."
14. Cohen et al., *Energy Down the Drain.*

Chapter 6. Constraints

Epigraph: *Poor Richard's Almanack,* January 1746, in the *Oxford Dictionary of American Quotations,* 2nd edition, edited by Hugh Rawson and Margaret Miner (New York: Oxford University Press, 2006), p. 68.

1. "European Heat Wave: August 16, 2003," *Earth Observatory,* National Aeronautics and Space Administration, image by Reto Stockli and Robert Simmon, based on data provided by the MODIS Land Science Team (http://earthobservatory.nasa.gov/IOTD/view.php?id=3714, accessed November 17, 2012).
2. Jean-Marie Robine et al., "Death Toll Exceeded 70,000 in Europe During the Summer of 2003," *Comptes Rendus Biologies* 331 (2008); M. Poumadère, C. Mays, S. L. Mer, and R. Blong, "The 2003 Heat Wave in France: Dangerous Climate Change Here and Now," *Risk Analysis* 25 (2005); and P. Lagadec, "Understanding the French 2003 Heat Wave Experience: Beyond the Heat, a Multi-Layered Challenge," *Journal of Contingencies and Crisis Management* 12, no. 4 (December 2004).
3. Julio Godoy, "Environment: European Heat Wave Shows Limits of Nuclear Energy," Inter Press Service News Agency, July 27, 2006 (www.ipsnews.net/2006/07/environment-heat-wave-shows-limits-of-nuclear-energy, accessed September 16, 2015).
4. H. Forster and J. Lilliestam, "Modeling Thermoelectric Power Generation in View of Climate Change," *Regional Environmental Change* 10 (2010); M. Poumadère, C. Mays, S. L. Mer, and R. Blong, "The 2003 Heat Wave in France:

Dangerous Climate Change Here and Now," *Risk Analysis* 25 (2005); M. Hightower and S. A. Pierce, "The Energy Challenge," *Nature,* March 20, 2008.

5. Illinois Environmental Protection Agency, "Illinois EPA Grants Exelon Quad Cities Station Provisional Variance from Discharge Requirements," Press Release, May 25, 2012.
6. Associated Press, "Conn. Nuclear Plant Unit Closed Due to Too-Warm Seawater Reopens," *Christian Science Monitor,* August 27, 2012.
7. "Outages and Curtailments During the Southwest Cold Weather Event of February 1–5, 2011," technical report, Federal Energy Regulatory Commission, August 2011; E. Souder, S. Gwynne, and G. Jacobson, "Freeze Knocked Out Coal Plants and Natural Gas Supplies, Leading to Blackouts," *Dallas Morning News,* February 6, 2011; E. Souder, G. Jacobson, and S. Gwynne, "Texas Electric Grid Operator's Rolling Blackouts During Freeze Bring Political Scrutiny," *Dallas Morning News,* February 12, 2011.
8. K. Galbraith, "The Rolling Chain of Events Behind Texas Blackouts," *Texas Tribune,* February 3, 2011.
9. "Outages and Curtailments During the Southwest Cold Weather Event."
10. "A Guide to the 2011 Texas Blackouts," Energy and Environment Reporting for Texas, *State Impact: A Reporting Project of NPR Member Stations* (http://stateimpact.npr.org/texas/tag/2011-blackouts, accessed November 26, 2012).
11. "Outages and Curtailments During the Southwest Cold Weather Event."
12. "In Pictures: Niagara Falls Freeze over on U.S. Side," *Telegraph,* January 10, 2014.
13. Christie Bleck, "Coal Shortage: WE Energies Buys BLP Coal After Heavy Ice on Lake Delays Ships," *Marquette Mining Journal,* 2014.
14. Rob Thomas, "Where's the Salt?" *Snow Magazine,* September 15, 2014; Brian Resnick, "America Is Running Low on Road Salt. Thanks, Winter," *National Journal,* February 5, 2014; "Midwest Experiences Propane Shortage Due to High Demand," WFIU, National Public Radio, January 24, 2014; Amy Wilson, "As Temps Dip, Propane Prices Heat Up," *Carthage Press,* January 21, 2014; "Propane Use for Crop Drying Depends on Weather and Corn Markets as Well as Crop Size," *Today in Energy,* Energy Information Administration, U.S. Department of Energy, October 2, 2014.
15. J. Gertner, "The Future Is Drying Up," *New York Times,* October 21, 2007; F. Proteger, "The Uruguay River, Its Dams, and Its People Are Running Out of Water," *International Rivers,* February 1, 2008; Scripps Institute of Oceanography, "Lake Mead Could Be Dry by 2021," Press Release (http://scrippsnews.ucsd.edu/Releases/?releaseID=876, February 12, 2008).

16. "Electricity Monthly Update," April 2012, technical report, Energy Information Administration, U.S. Department of Energy, June 2012.
17. S. Gottipati, "India Endures World's Largest Blackout," *New York Times,* July 31, 2012; J. Yardley and G. Harris, "Second Day of Power Failures Cripples Wide Swath of India," *New York Times,* July 31, 2012; *Hindustan Times,* "Power Grid Failures: FAQs" (www.hindustantimes.com/India-news/NewDelhi/Power-grid-failure-FAQs/Article1-905428.aspx, accessed January 4, 2015); BBC, "Power Cut Causes Major Disruption in Northern India" (www.bbc.co.uk/news/world-asia-india-19043972, July 30, 2012, accessed January 4, 2015).
18. Associated Press, "Drought Could Shut Down Nuclear Power Plants," MSNBC, January 23, 2008.
19. "Water Levels Are Rising Here on Lake Lanier!" (www.lakelanier.com/about/water-levels, accessed November 22, 2012); L. Mungin, "Two Off-Line Power Plants Help Region Hit Water Goal," *Atlanta Journal-Constitution,* December 20, 2007.
20. M. Hightower and S. A. Pierce, "The Energy Challenge," *Nature,* March 20, 2008; P. N. Spotts, "Wanna Save Water? Plunge into the Car Pool," *Christian Science Monitor,* November 3, 2008; Ministère de l'Ecologie, du Développement Durable et de l'Energie, "La description du parc électronucléaire Français" (www.developpement-durable.gouv.fr/La-production-d-electricite.html, updated March 10, 2011, accessed November 22, 2012).
21. J. Keen, "Low River May Paralyze Commerce," *USA Today,* November 2012.
22. A. G. Sulzberger and Matthew L. Wald, "Flooding Brings Worries over Two Nuclear Plants," *New York Times,* June 20, 2011; Associated Press, "Flood Berm Collapses at Nebraska Nuclear Plant," June 26, 2011.
23. J. Paul, "Experts: Ethanol's Water Demands a Concern," *Journal-Advocate,* Associated Press, June 19, 2006.
24. M. Lee, "Parched Texans Impose Water-Use Limits for Fracking Gas Wells," *Business Week,* October 6, 2011.
25. Ibid.
26. M. A. Cook and M. E. Webber, "Food, Fracking, and Freshwater: The Potential for Markets and Cross-Sectoral Investments to Enable Water Conservation," *Water* (in review).
27. B. D. Lutz, A. N. Lewis, and M. W. Doyle, "Generation, Transport, and Disposal of Wastewater Associated with Marcellus Shale Gas Development," *Water Resources Research* 49 (2013).

28. Y. Kuwayama, S. Olmstead, and A. Krupnick, "Water Quality and Quantity Impacts of Hydraulic Fracturing," *Current Sustainable Renewable Energy Reports,* 2015.
29. Lutz et al., "Generation, Transport, and Disposal of Wastewater."
30. Ibid.
31. W.-Y. Kim, "Induced Seismicity Associated with Fluid Injection into a Deep Well in Youngstown, Ohio," *Journal of Geophysical Research Solid Earth* 118 (2013), doi:10.1002/jgrb.50247.
32. Kuwayama et al., "Water Quality and Quantity Impacts of Hydraulic Fracturing."
33. Dina Cappiello, Frank Bass, and Cain Burdeau, "AP Investigation: Ike Environmental Toll Apparent," *USA Today,* October 6, 2008.
34. Michael Schwirtz, "Sewage Flows After Storm Expose Flaws in System," *New York Times,* November 29, 2012; "Raw Sewage Still Plagues Long Island Homes Five Weeks After Sandy," *CBS New York,* December 4, 2012 (http://newyork.cbslocal.com/2012/12/04/raw-sewage-still-plagues-long-island-homes-5-weeks-after-sandy, accessed December 8, 2012).

Chapter 7. Trends

1. Jill Boberg, *Liquid Assets* (Santa Monica: Rand Corporation, 2005); Peter H. Gleick, editor, *Water in Crisis: A Guide to the World's Fresh Water Resources* (New York: Oxford University Press, 1993); Energy Information Administration and International Energy Agency reports.
2. U.S. Geological Survey, "Water-Level Changes in the High Plains Aquifer, Predevelopment to 2007, 2005–06, and 2006–07," technical report, 2009, Scientific Investigations Report 20095019.
3. Boberg, *Liquid Assets;* C. J. Vörösmarty et al., "Global Threats to Human Water Security and River Biodiversity," *Nature,* September 30, 2010.
4. Y. Pokhrel, N. Hanasaki, P.-F. Yeh, T. Yamada, S. Kanae, and T. Oki, "Model Estimates of Sea-Level Change Due to Anthropogenic Impacts on Terrestrial Water Storage," *Nature Geoscience* 5 (June 2012).
5. Kai Ryssdal, "California's Snowpack Has Run Out," *Marketplace,* APM Radio, May 29, 2015.
6. U.S. National Oceanic and Atmospheric Administration, "Global Climate Change Impacts in the United States" (New York: Cambridge University Press, 2009).
7. Felicity Barringer, "Troubling Interdependency of Water and Power," *New York Times,* April 22, 2015.

8. B. Boehlert et al., "Climate Change Impacts and Greenhouse Gas Mitigation Effects on U.S. Hydropower Generation," *Environmental Research Letters,* 2015.
9. C. W. King and M. E. Webber, "Water Intensity of Transportation," *Environmental Science and Technology* 42 (September 24, 2008).
10. C. W. King, M. E. Webber, and I. J. Duncan, "The Water Needs for LDV Transportation in the United States," *Energy Policy* 38, no. 2 (2010).
11. David Sedlak, *Water 4.0: The Past, Present and Future of the World's Most Vital Resource* (New Haven: Yale University Press, 2014).
12. K. M. Twomey, A. S. Stillwell, and M. E. Webber, "The Unintended Energy Impacts of Increased Nitrate Contamination from Biofuels Production," *Journal of Environmental Monitoring* 12 (2010).
13. Lutz et al., "Generation, Transport, and Disposal of Wastewater."
14. Peter Gleick et al., *The World's Water,* volume 8 (Washington, D.C.: Island, 2014), chapter 7.
15. "South–North Water Transfer Project," International Rivers (www.internationalrivers.org/campaigns/south-north-water-transfer-project, accessed January 4, 2015).
16. "The 2012 State Water Plan," technical report, Texas Water Development Board, 2012.
17. Susan Berfield, "There Will Be Water," *Bloomberg Businessweek,* June 11, 2008.
18. David Mildenberg, "Pickens Water-to-Riches Dream Unravels as 11 Texas Cities Scoop Up Rights," *Bloomberg News,* July 13, 2011.
19. Global Water Intelligence, *Desalination Markets 2010: Global Forecast and Analysis,* technical report, Desal Data, 2010; International Desalination Association, "Desalination by the Numbers" (http://idadesal.org/desalination-101/desalination-by-the-numbers, accessed August 15, 2015).
20. Isabel Kershner, "Aided by the Sea, Israel Overcomes an Old Foe: Drought," *New York Times,* May 29, 2015.
21. Global Water Intelligence, *Desalination Markets 2010.*
22. A. S. Stillwell, C. W. King, and M. E. Webber, "Desalination and Long-Haul Water Transfer as a Water Supply for Dallas, Texas: A Case Study of the Energy-Water Nexus in Texas," *Texas Water Journal* 1, no. 1 (September 2010).

Chapter 8. Technical Solutions

1. Emily A. Grubert, Fred C. Beach, and Michael E. Webber, "Can Switching Fuels Save Water? A Life Cycle Quantification of Freshwater Consumption

for Texas Coal- and Natural Gas–Fired Electricity," *Environmental Research Letters* 7 (2012).

2. A. S. Stillwell and M. E. Webber, "Water Conservation and Reuse: A Case Study of the Energy-Water Nexus in Texas," conference paper, *World Environmental and Water Resources Congress, 2010,* Providence, Rhode Island.
3. H. Li, S.-H. Chien, M.-K. Hsieh, D. A. Dzombak, and R. D. Vidic, "Escalating Water Demand for Energy Production and the Potential for Use of Treated Municipal Wastewater," *Environmental Science and Technology* 45 (2011): 4195–4200.
4. A. S. Stillwell, K. M. Twomey, R. Osborne, D. M. Greene, D. W. Pedersen, and M. E. Webber, "An Integrated Energy, Carbon, Water, and Economic Analysis of Reclaimed Water Use in Urban Settings: A Case Study of Austin, Texas," *Journal of Water Reuse and Desalination,* 2011.
5. "Water Reuse: Potential for Expanding the Nation's Water Supply Through Reuse of Municipal Wastewater," National Research Council, 2012.
6. Ibid.
7. U.S. Geological Survey, "Alternative Water Use," presentation by Melissa Harris to the Third Energy-Water Nexus Partnership Meeting, November 29, 2012; "Use of Reclaimed Water for Power Plant Cooling," technical report ANL/EVS/R-07/3, Argonne National Laboratory for the National Energy Technology Lab, U.S. Department of Energy, August 2007.
8. "Palo Verde Nuclear Generating Station Water Reclamation Facility," technical report, Electric Power Research Institute, June 26, 2008.
9. Killington Ski Resort, "Snow You Can Count On" (www.killington.com/winter/mountain/mountain_info/guarantee, accessed November 25, 2012).
10. K. D. Lorentz, "Facing Obstacles: History Lessons After an Important Apology Bring Hope for Progress," *Mountain Times,* May 23, 2012.
11. Ibid.
12. L. Macmillan, "Resort's Snow Won't Be Pure This Year; It'll Be Sewage," *New York Times,* September 26, 2012.
13. "Use of Reclaimed Water for Power Plant Cooling."
14. U.S. Geological Survey, "Alternative Water Use."
15. P. F. Scholander, "How Mangroves Desalinate Seawater," *Physiologia Plantarum* 21 (1968): 251–261.
16. Sam Lemonick, "Harvesting Fog Could Bring Water to Millions," *Earth Magazine,* February 2014.
17. A. S. Stillwell and M. E. Webber, "A Novel Methodology for Evaluating Economic Feasibility of Low-Water Cooling Technology Retrofits at Power Plants," *Water Policy* 15 (2013).

18. Ibid.; "Oyster Creek to Close 10 Years Early, in 2019," *Asbury Park Press* (Neptune, N.J.), December 8, 2010.
19. Grubert, Beach, and Webber, "Can Switching Fuels Save Water?"
20. Kate Galbraith, "Waterless Fracking Makes Headway in Texas, Slowly," *Texas Tribune,* March 27, 2013.
21. U.S. Department of Energy, "Variable Speed Pumping: A Guide to Successful Applications," technical report DOE/GO-102004-1913, May 2004.
22. Ibid.
23. E. M. Keys and M. E. Webber, "Variable Speed Drives for Power Factor Correction in the Water Sector," Fifth International Symposium on Power Electronics for Distributed Generation (PEDG) Systems, IEEE, June 24–27, 2014, Galway, Ireland.
24. K. T. Sanders and M. E. Webber, "Evaluating the Energy and Carbon Dioxide Emissions Impacts of Shifts in Residential Water Heating in the United States," *Energy,* 2015.
25. Eric Lajoie-Mazenc, "Which Technologies for Hot Water in France for 21st Century?" ACEEE Hot Water Forum, Berkeley, California, May 2012.
26. Mark Tran, "WaterWheel to Ease Burden on Women," *Guardian,* December 29, 2013.
27. A. S. Stillwell, D. C. Hoppock, and M. E. Webber, "Energy Recovery from Wastewater Treatment Plants in the United States: A Case Study of the Energy-Water Nexus," *Sustainability* 2 (2010), no. 4 (special issue: Energy Policy and Sustainability).
28. Y. R. Glazer, J. B. Kjellsson, K. T. Sanders, and M. E. Webber, "The Potential for Using Energy from Flared Gas for On-Site Hydraulic Fracturing Wastewater Treatment in Texas," *Environmental Science and Technology Letters,* 2014.
29. M. E. Clayton, A. S. Stillwell, and M. E. Webber, "Implementation of Brackish Groundwater Desalination Using Wind-Generated Electricity: A Case Study of the Energy-Water Nexus in Texas," *Sustainability* 6 (2014), no. 2 (special issue: The Energy-Sustainability Nexus).
30. M. E. Clayton, J. B. Kjellsson, and M. E. Webber, "Wind-Solar-Desalination: How Integrated Systems Can Solve Our Water and Energy Issues," *Earth,* November 2014.
31. E. A. Grubert, A. S. Stillwell, and M. E. Webber, "Where Does Solar-Aided Seawater Desalination Make Sense? A Method for Identifying Sustainable Sites," *Desalination* 339 (2014).
32. Communication from the Department of Economic Development, United Arab Emirates, 2010.

Chapter 9. Nontechnical Solutions

1. Daniel M. Kammen and Gregory F. Nemet, "Reversing the Incredible Shrinking Energy R&D Budget," *Issues in Science and Technology,* Fall 2005.
2. John F. Kennedy, "The President's News Conference," April 12, 1961, quoted online by Gerhard Peters and John T. Woolley, *The American Presidency Project,* University of California, Santa Barbara (www.presidency.ucsb.edu/ws/?pid=8055).
3. S. R. Kirshenbaum and M. E. Webber, "Time for Another Giant Leap for Mankind," *Issues in Science and Technology,* Spring 2012.
4. M. E. Webber, "Breaking the Energy Barrier," *Earth,* September 2009.
5. "Hidden Costs of Energy: Unpriced Consequences of Energy Production and Use," technical report, National Research Council of the National Academies, 2010.
6. P. Epstein et al., "Full Cost Accounting for the Life Cycle of Coal," *Annals of the New York Academy of Sciences: Ecological Economics Reviews,* 2011.
7. M. A. Cook and M. E. Webber, "Food, Fracking, and Freshwater: The Potential for Markets and Cross-Sectoral Investments to Enable Water Conservation," *Water* (in review).
8. National Research Council, "Real Prospects for Energy Efficiency," 2010.
9. Trey Thoelcke and Michael B. Sauter, "The Nine Most Successful Retail Stores in the USA," *USA Today,* November 18, 2012.
10. Finish Line car wash, "Did You Know?" (www.finishlinewash.com/gpage1.html, accessed October 18, 2012).

Acknowledgments

Many important thinkers and researchers in my research group performed the bulk of the research presented here. This book benefited extensively from the scientific advances of Ashlynn Stillwell, Carey King, and Kelly Twomey Sanders, whose contributions are featured throughout based on various papers and reports we wrote in collaboration. They represent the core of my early energy-water-nexus team within my research group at the University of Texas at Austin, and I am very grateful for their tireless diligence and scientific rigor. This project and program of work would not have been possible without them, and their contributions to this manuscript cannot be overstated.

Carey King's contributions to the water intensity of transportation fuels and water needs for power plants are significant. Kelly Sanders's look at the energy intensity of water was groundbreaking. In particular I lean very heavily on the extensive body of research and teaching materials fostered by Ashlynn Stillwell. The dozen-plus papers and reports we authored together and the courses we taught together while she was a graduate student in my group form the backbone for many sections in this manuscript.

Elements from my collaborative projects with Sheril Kirshenbaum appear in the sections on water research and development and on energy, water, and women. Plus, many others contributed important concepts or pieces of knowledge, including Colin Beal, Mary Clayton, Margaret Cook, Todd Davidson, Yael Glazer, Gary Gold, Emily Grubert, David Hoppock, Erin Keys, Jill Kjellsson, Charlie Upshaw, and others. Other early collaborators included Ian Duncan from the Bureau of Economic Geology at UT Austin and Amy Hardberger from St. Mary's School of Law. Along the way, I also learned a lot from Bridget Scanlon and J. P. Nicot, whose work at the Bureau of Economic Geology is authoritative. I had the opportunity to collaborate with several insightful researchers through a large-scale study of the energy-water nexus conducted by the Union of Concerned Scientists, including John Rogers, Peter Frumhoff, and others, which was helpful for my thinking on this topic. I also had the good fortune to collaborate with several experts at the U.S. Department of Energy, including Vince Tidwell, Mike Hightower, and Robin Newmark.

Marilu Hastings at the Cynthia and George Mitchell Foundation had the foresight to fund many of the research projects that built up this body of work. Her patient sponsorship, thought leadership, and friendship along with the foundation's steady support has helped bring the importance of this topic to the forefront of policy discussions.

I learned important insights about energy for water from Gary Klein, sustainable business models from Dave Allen, and the shift from cost-based to value-based capitalism from Joe Stanislaw. Jeff Phillips pulled together the art, illustrations, and graphics, which improved the book immensely. Coleman Tharpe helped with the communications plan. I thank all of them for their efforts and important contributions to this work.

I especially want to thank Marianne Shivers Gonzalez for her help with the tedious and critically important task of reviewing the manuscript at multiple stages in its development. Every writer needs an editor, and Marianne bravely and very capably took on that responsibility. In that spirit, I also want to thank Melissa Chinchillo at Fletcher & Company and Jean Thomson Black at Yale University Press, who agreed to work with me on this project. Their suggestions and nudging were very valuable. The book's peer reviewers, including Kate Galbraith, whose writings on this topic are among the best, helped make the book more rigorous and interesting. I learned a great deal about the history of water from David Sedlak.

John and Kristen Schulz gave me clarity that the time had arrived to write this book and Aunt Debbie is the one who reignited my passion for this topic. Russell Gold, Peter Gleick, and Bill Brands all gave me encouragement and specific tips on how to be productive at writing. A mentoring session with Peter in 2012 got me into high gear; my frequent lunches with Russell demonstrated that there is light at the end of the tunnel; and Bill taught me the mechanics of how to be somebody who writes books instead of someone who says they will write books. Johnnie Johnson coached me through the process of writing, and Mark Fischetti worked with me on the article that launched the book.

I would also like to thank Dartmouth College for hosting me as a visiting associate professor during the fall of 2012, when the bulk of this manuscript was initially pulled together from the outline I had scribbled in 2005. I am also indebted to the Donald D. Harrington Faculty Fellows Program, whose support in 2014 granted me the time I needed to pull together the missing pieces. Last, I appreciate the support of my wife, who listened to me grumble for many years about how this book needed to be written, and gave me the support to get it done.

Index

Page numbers in italic type refer to illustrations